# Mathematics
# Entertainment
# for the Millions

# Problem Solving in Mathematics and Beyond

Print ISSN: 2591-7234
Online ISSN: 2591-7242

**Series Editor:** Dr. Alfred S. Posamentier
Distinguished Lecturer
New York City College of Technology - City University of New York

There are countless applications that would be considered problem solving in mathematics and beyond. One could even argue that most of mathematics in one way or another involves solving problems. However, this series is intended to be of interest to the general audience with the sole purpose of demonstrating the power and beauty of mathematics through clever problem-solving experiences.

Each of the books will be aimed at the general audience, which implies that the writing level will be such that it will not engulfed in technical language — rather the language will be simple everyday language so that the focus can remain on the content and not be distracted by unnecessarily sophiscated language. Again, the primary purpose of this series is to approach the topic of mathematics problem-solving in a most appealing and attractive way in order to win more of the general public to appreciate his most important subject rather than to fear it. At the same time we expect that professionals in the scientific community will also find these books attractive, as they will provide many entertaining surprises for the unsuspecting reader.

## Published

For the complete list of volumes in this series, please visit www.worldscientific.com/series/psmb

**Problem Solving in Mathematics and Beyond** Volume **18**

# Mathematics Entertainment for the Millions

Alfred S. Posamentier

City University of New York, USA

 **World Scientific**

NEW JERSEY · LONDON · SINGAPORE · BEIJING · SHANGHAI · HONG KONG · TAIPEI · CHENNAI · TOKYO

*Published by*

World Scientific Publishing Co. Pte. Ltd.

5 Toh Tuck Link, Singapore 596224

*USA office:* 27 Warren Street, Suite 401-402, Hackensack, NJ 07601

*UK office:* 57 Shelton Street, Covent Garden, London WC2H 9HE

**Library of Congress Cataloging-in-Publication Data**

Names: Posamentier, Alfred S., author.

Title: Mathematics entertainment for the millions /
    Alfred S. Posamentier, City University of New York, USA.

Description: New Jersey : World Scientific, [2020] | Series: Problem solving in
    mathematics and beyond, 2591-7234 ; volume 18 | Includes index.

Identifiers: LCCN 2020013734 | ISBN 9789811219900 (hardcover) |
    ISBN 9789811219283 (paperback) | ISBN 9789811219290 (ebook for institutions) |
    ISBN 9789811219306 (ebook for individuals)

Subjects: LCSH: Mathematical recreations. | Mathematics--Problems, exercises, etc.

Classification: LCC QA95 .P66 2020 | DDC 793.74--dc23

LC record available at https://lccn.loc.gov/2020013734

**British Library Cataloguing-in-Publication Data**

A catalogue record for this book is available from the British Library.

For any available supplementary material, please visit
https://www.worldscientific.com/worldscibooks/10.1142/11795#t=suppl

Desk Editors: V. Vishnu Mohan/Tan Rok Ting

Typeset by Stallion Press
Email: enquiries@stallionpress.com

# About the Author

 **Alfred S. Posamentier** is currently Distinguished Lecturer at New York City College of Technology of the City University of New York. Prior to that he was Executive Director for Internationalization and Funded Programs at Long Island University, New York. This was preceded by 5 years as Dean of the School of Education and Professor of Mathematics Education at Mercy College, New York. Before that he was for 40 years at The City College of the City University of New York, at which he is now Professor Emeritus of Mathematics Education at and Dean Emeritus of the School of Education. He is the author and co-author of more than 75 mathematics books for teachers, secondary and elementary school students, as well as the general readership. Dr. Posamentier is also a frequent commentator in newspapers and journals on topics related to education.

After completing his B.A. degree in mathematics at Hunter College of the City University of New York, he took a position as a teacher of mathematics at Theodore Roosevelt High School (Bronx, New York), where he focused his attention on improving the students' problem-solving skills and at the same time enriching their instruction far beyond what the traditional textbooks offered. During his six-year tenure there, he also developed the school's first mathematics teams (both at the junior and senior level). He is still involved in working with mathematics teachers and supervisors, nationally and internationally, to help them maximize their effectiveness.

Immediately upon joining the faculty of the City College of New York in 1970 (after having received his master's degree there in 1966), he began to develop in-service courses for secondary school mathematics teachers, including such special areas as recreational mathematics and problem solving in mathematics. As Dean of the City College School of Education for 10 years, his scope of interest in educational issues covered the full gamut educational issues. During his tenure as dean he took the School from the bottom of the New York State rankings to the top with a perfect NCATE accreditation assessment in 2009. Posamentier repeated this successful transition at Mercy College, which was then the only college to have received both NCATE and TEAC accreditation simultaneously.

In 1973, Dr. Posamentier received his Ph.D. from Fordham University (New York) in mathematics education and has since extended his reputation in mathematics education to Europe. He has been visiting professor at several European universities in Austria, England, Germany, Czech Republic, Turkey and Poland. In 1990, he was Fulbright Professor at the University of Vienna.

In 1989, he was awarded an Honorary Fellow position at the South Bank University (London, England). In recognition of his outstanding teaching, the City College Alumni Association named him Educator of the Year in 1994, and in 2009. New York City had the day, May 1, 1994, named in his honor by the President of the New York City Council. In 1994, he was also awarded the *Das Grosse Ehrenzeichen für Verdienste um die Republik* Österreich, (Grand Medal of Honor from the Republic of Austria), and in 1999, upon approval of Parliament, the President of the Republic of Austria awarded him the title of University Professor of Austria. In 2003, he was awarded the title of *Ehrenbürgerschaft* (Honorary Fellow) of the Vienna University of Technology, and in 2004 was awarded the *Österreichisches Ehrenkreuz für Wissenschaft & Kunst 1. Klasse* (Austrian Cross of Honor for Arts and Science, First Class) from the President of the Republic of Austria. In 2005, he was inducted into the Hunter College

Alumni Hall of Fame, and in 2006 he was awarded the prestigious Townsend Harris Medal by the City College Alumni Association. He was inducted into the New York State Mathematics Educator's Hall of Fame in 2009, and in 2010 he was awarded the coveted Christian-Peter-Beuth Prize from the Technische Fachhochschule — Berlin. In 2017, Posamentier was awarded *Summa Cum Laude nemmine discrepante,* by the Fundacion Sebastian, A.C., Mexico City, Mexico.

He has taken on numerous important leadership positions in mathematics education locally. He was a member of the New York State Education Commissioner's Blue Ribbon Panel on the Math-A Regents Exams, and the Commissioner's Mathematics Standards Committee, which redefined the Mathematics Standards for New York State, and he also served on the New York City schools' Chancellor's Math Advisory Panel.

Dr. Posamentier is still a leading commentator on educational issues and continues his long time passion of seeking ways to make mathematics interesting to both teachers, students and the general public — as can be seen from some of his more recent books, *Math Makers: The Lives and Works of 50 Famous Mathematicians* (Prometheus Books, 2020), *Understanding Mathematics Through Problem Solving* (World Scientific Publishing, 2020); *The Psychology of Problem Solving: The Background to Successful Mathematics Thinking* (World Scientific Publishing, 2020); *Solving Problems in Our Spatial World* (World Scientific Publishing, 2019); *Tools to Help Your Children Learn Math: Strategies, Curiosities, and Stories to Make Math Fund or Parents and Children* (World Scientific Publishing, 2019); *The Mathematics of Everyday Life* (Prometheus, 2018), *The Joy of Mathematics* (Prometheus Books, 2017); *Strategy Games to Enhance Problem — Solving Ability in Mathematics* (World Scientific Publishing, 2017); *The Circle: A Mathematical Exploration Beyond the Line* (Prometheus Books, 2016), *Effective Techniques to Motivate Mathematics Instruction*, 2nd Ed. (Routledge, 2016); *Problem — Solving Strategies in Mathematics* (World Scientific Publishing, 2015) *Numbers: There Tales, Types, and Treasures* (Prometheus Books, 2015), *Mathematical Curiosities* (Prometheus Books, 2014); *Magnificent Mistakes in Mathematics* (Prometheus Books, 2013), 100 *Commonly*

*Asked Questions in Math Class: Answers that Promote Mathematical Understanding, Grades 6-12* (Corwin, 2013), *What Successful Math Teachers do - Grades 6-12* (Corwin, 2013), *The Secrets of Triangles: A Mathematical Journey* (Prometheus Books, 2012), *The Glorious Golden Ratio* (Prometheus Books, 2012), *The Art of Motivating Students for Mathematics Instruction* (McGraw-Hill, 2011), *The Pythagorean Theorem: Its Power and Glory* (Prometheus, 2010), *Teaching Secondary Mathematics: Techniques and Enrichment Units,* 9th Ed. (Pearson, 2015), *Mathematical Amazements and Surprises: Fascinating Figures and Noteworthy Numbers* (Prometheus, 2009), *Problem Solving in Mathematics: Grades 3-6: Powerful Strategies to Deepen Understanding* (Corwin, 2009), *Problem-Solving Strategies for Efficient and Elegant Solutions, Grades 6-12* (Corwin, 2008), *The Fabulous Fibonacci Numbers* (Prometheus Books, 2007), *Progress in Mathematics K-9* textbook series (Sadlier-Oxford, 2006–2009), *What successful Math Teacher Do: Grades K-5* (Corwin 2007), *Exemplary Practices for Secondary Math Teachers* (ASCD, 2007), *101+ Great Ideas to Introduce Key Concepts in Mathematics* (Corwin, 2006), π, *A Biography of the World's Most Mysterious Number* (Prometheus Books, 2004), *Math Wonders: To Inspire Teachers and Students* (ASCD, 2003), and *Math Charmers: Tantalizing Tidbits for the Mind* (Prometheus Books, 2003).

# Contents

# Introduction

To entertain folks is rather easy. One can tell a joke, one can tell an interesting story, one can show some interesting photographs, one might even want to sing a song. These are all common ways to entertain the general public. However, unexpected as it might seem at this point, one can also entertain folks with unusual and yet amazing aspects of mathematics. Some of these can be very simple and can be done without even writing anything. Others will be considered awesome and can lead the audience to truly appreciate mathematics, as perhaps, they have missed doing so during their school days. It is unfortunate that most of the school instruction in mathematics is guided by a firm curriculum and controlled by regular testing. Furthermore, some school districts rate teachers' instructional performance by how well their students perform on the standardized tests. This results in a rather firm instructional program that is often referred to as "teaching to the test." It is unfortunate that such an instructional program leaves little room to show some of the beautiful aspects of mathematics. Many of these can be presented in a light-hearted fashion, which could be quite entertaining. Before we embark on our journey through the entertaining aspects of mathematics, it is important to understand what makes for this kind of entertainment.

When one polls a general audience, it is not uncommon to find that the majority of people will be proud to say that they were not particularly strong in mathematics during their school days, as though this is a badge of honor to carry into the future. Yet, it is hoped that our journey through entertaining aspects of mathematics will change this perception of this most important subject. For the most part, to be entertaining the presentation must be lighthearted and brief and not be built on too much previous mathematical experience. One might even say that "table talk" could be a good forum for such entertaining aspects.

For example, if you are talking with friends and want to impress them with unusual cleverness in mathematics, you might want to show them how you can mentally multiply a two-digit number by 11. All you need to do is to tell them to add the two digits and take that sum and place it between these two digits. For example, $34 \times 11$ would require adding $3 + 4 = 7$ and placing the 7 between the 3 and the 4 to get 374. An immediate question that might arise is "What do you do if the sum of the two digits is greater than 9?" Here, we simply carry the 1 and add it to the original tens digit. As is the case when we multiply $78 \times 11$, where $7 + 8 = 15$ and we place the 5 between the 7 and 8 and add the 1 to the 7, so that the answer is 858. We will cover this topic in greater detail later in the book when we extend multiplying by 11 to numbers consisting of more than two digits. Again, when doing any entertaining, the timing is important and the mode of presentation, which we will try to address as we present these many aspects of mathematical entertainment.

There are times, when an unusual mathematical curiosity can be shown to an audience to get response of "disbelief." For example, where you take any size number and then the number with the digits reversed. When you subtract the two numbers, the result will always be divisible by 9. Rather than to rush through this example, it is always wise to allow the audience to try it with their own selected number. We can take, for example, the number 1357 and the number with the digits reversed is 7531. The difference of these two numbers is 6174, which equals $9 \times 686$. By the way, with this random example we just happened to stumble on the number 6174 for which there are a host of entertaining wonders attached which we will explore later in the book.

Throughout this book, we will be providing a plethora of mathematical amusements from the fields of geometry, algebra, probability, logic, and others such as the arithmetic ones we just encountered. There are some mathematics entertainments that require some logical thinking — and can be explained through algebra or other traditional aspects of mathematics. However, above all they are easy to follow, have truly unexpected results and are genuinely entertaining. As an illustration one such can be done with coins and will show you how some clever reasoning along with very elementary algebraic knowledge will help you sort out this unexpected result. Suppose your friend is seated at a table in a dark room. On the table, there are 12 pennies, 5 of which are heads up and 7 are tails up. She knows where the coins are, so she can mix them by sliding around the coins, but because the room is dark, she will not know if the coins that she is touching were originally heads up or tails up. You now ask her to separate the coins into two piles of 5 and 7 coins and then flip all the coins in the 5-coin group. To everyone's amazement, when the lights are turned on there will be an equal number of heads in each of the two piles. Your friend's first reaction is "you must be kidding!" How can anyone do this task without seeing which coins are heads up or tails up? The solution will surely enlighten the friend and at the same time illustrate how algebraic symbols can help understanding.

Let's now look at the explanation of this surprise result. This is where a clever (yet incredibly simple) use of algebra will be the key to explaining the unexpected outcome. Lets "cut to the chase." The 12 coins have 5 with heads up and 7 with tails up. Without being able to look at the coins, she separated the coins into two piles, of 5 and 7 coins each. Then she flipped over the coins in the smaller pile of 5 coins. Then both piles had the same number of heads!

Well, this is where a little algebra helps us understand what was actually done. Let's say that when she separated the coins in the dark room, $h$ heads will end up in the 7-coin pile. Then the other pile, the 5-coin pile, will have $5 - h$ heads. To get the number of tails in the 5-coin pile, we subtract the number of heads $(5 - h)$ from the total number of coins in the pile, 5, to get: $5 - (5 - h) = h$ tails.

| 5-coin pile | 7-coin pile |
|---|---|
| $5 - h$ heads | $h$ heads |
| $5 - (5 - h)$ tails<br>$= h$ tails | |

When she flips all the coins in the smaller pile (the 5-coin pile), the $(5 - h)$ heads become tails and the $h$ tails become heads. Now each pile contains $h$ heads!

**The piles after flipping the coins in the smaller pile**

| 5-coin pile | 7-coin pile |
|---|---|
| $5 - h$ tails<br>$h$ heads | $h$ heads |

This absolutely surprising result will show you how the simplest algebra can explain an entertaining aspect of mathematics.

With these brief examples, we hope to have whet your appetite for the many unexpected and counterintuitive entertainments that the field of mathematics can offer. Join us now as we journey through these recreational aspects of mathematics.

# Chapter 1

# Arithmetical Entertainments

Most people recall the concept of arithmetic as the subject that they learned in elementary school. For them, this concept consists largely of the four basic operations: addition, subtraction, multiplication, and division. Throughout the course of their schooling, this extended to doing computation with fractions and perhaps square root extraction. In our constantly-developing technological world, when these tasks are required today, most folks revert to the calculator to do the arithmetic calculations. When the knowledge or concept of arithmetic is limited to its barest form of usefulness, much of the beauty of mathematics gets lost. Actually, arithmetic lends itself beautifully to some very simple entertaining aspects. When you get to the end of this chapter, you will find what some might consider a mind-blowing phenomenon in arithmetic. However, let's start off with an easy little puzzle-type question that requires a little bit of arithmetic insight.

## An Easy One for Starters

There are times when a very simple question can prove a bit challenging. Consider the dartboard shown in Figure 1.1, where the values of each ring are given. What is the least number of arrows that need to be shot at this dartboard in order to get an exact total of 100?

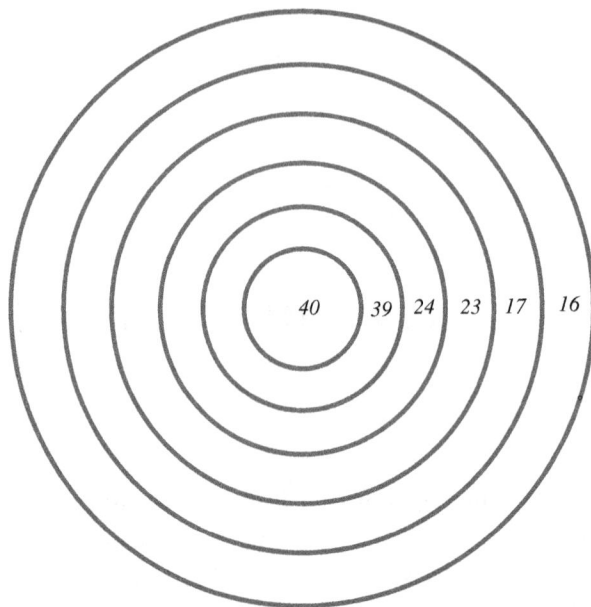

**Figure 1.1**

As your audience begins to try various combinations and before too much frustration arises, you might relieve them by telling them that the 6 arrows required would have to hit the number 17 four times and hit the number 16 twice, since $(4 \times 17) + (2 \times 16) = 68 + 32 = 100$.

Now that your audience is an expert on this type of arithmetic, have them consider how you can form the number 100 most efficiently with the numbers 11, 13, 31, 33, 42, 44, 46. And the answer is: $11 + 11 + 13 + 13 + 13 + 13 + 13 + 13 = 100$. You might want to develop other arithmetic entertainments of the sort if your audience finds it enjoyable.

## A Puzzling Question

As a further warm-up to get some insights into arithmetic and its little secrets, we consider the following puzzling question:

Find the fraction that when its double is added to its half and then is multiplied by the original fraction, the result is the original fraction.

To solve this, we simply need to realize that if we multiply the result of the first few arithmetic requirements by the original number, then that result would need to be 1. Therefore, we need to start off with a fraction that when it is double and added to half of it will yield 1. That fraction is $\frac{2}{5}$, since when multiplying by 2, we get $\frac{4}{5}$ and then taking half of $\frac{2}{5}$, gives us $\frac{1}{5}$, when added to $\frac{4}{5}$, gives us $\frac{4}{5}+\frac{1}{5}=\frac{5}{5}=1$. Therefore, when we multiply $1 \times \frac{2}{5}=\frac{2}{5}$, which was our desired result.

Another warm-up starter that might be fun would be to find three equal fractions using the numbers from 1 to 9 exactly once. After you have given your audience a bit of time to craft some possibilities, here are a few that will allow you to be properly prepared:

$$\frac{2}{4}=\frac{3}{6}=\frac{79}{158}$$

$$\frac{3}{6}=\frac{7}{14}=\frac{29}{58}$$

$$\frac{3}{6}=\frac{9}{18}=\frac{27}{54}$$

$$\frac{2}{6}=\frac{3}{9}=\frac{58}{174}$$

Such problems can make a very entertaining — and sometimes challenging — time for the audience. Continue with us now on a journey through lots of unusual, surprising, and entertaining aspects that arithmetic provides.

## Some Cute Arithmetic Tidbits

Often cute little tidbits, which are easy to present and have surprising results can further motivate an audience. Although each of the following brief curiosities nicely motivate the audience, it should be remembered — now and throughout the entire book — that the style in which the presentation is made, that is, the enthusiasm that the presenter harbors, is key to appropriately entertaining the audience. So, let's get started and consider the following short curiosities:

- You might want to begin by giving your audience a cute little challenge. Ask them to determine which symbol can be placed between 2 and 3 so that the result will be a number greater than 2 and less than 3. They will probably try to use the typical arithmetic operations, such as $\times + - \div$ and find that none of them produces a number between 2 and 3. They will be surprised with the simple response that a decimal point symbol would have solved the problem by yielding 2.3.

- It is well known largely because of the Pythagorean theorem that $3^2 + 4^2 = 5^2$, however, surprisingly, it is not so well known that $3^3 + 4^3 + 5^3 = 6^3$.

- While considering the previously mentioned Pythagorean triple, namely, (3, 4, 5) since $3^2 + 4^2 = 5^2$, we notice that the first two numbers are consecutive. Challenge your audience to see if they can find other Pythagorean triples where the first two numbers are also consecutive. The next one would be (20, 21, 29), since $20^2 + 21^2 = 29^2$. So that you are properly prepared; here are the next such triples (119, 120, 169) and (696, 697, 985). Another one that meets this criterion is (803760, 803761, 1136689). It would be quite challenging (and perhaps even entertaining) to find other such triples.

  To find Pythagorean triples where the second two members are consecutive is much simpler. Here are just a few to consider (3, 4, 5), (5, 12, 13), (7, 24, 25), (9, 40, 41), (11, 60, 61), (13, 84, 85), (15, 112, 113), and on it goes without end.

- The sum of all the factors of the square number 81 is equal to another square number 121. We can show that as $1 + 3 + 9 + 27 + 81 = 121$.

- The numbers 139 and 149 are the smallest consecutive prime numbers whose difference is 10.
- We know that $30 - 33 = -3$. How can we move one digit so that the answer becomes $+3$? The answer is rather simple: $30 - 3^3 = 30 - 27 = +3$. Cute!
- We can admire the number 48 as having a very peculiar property. If we add 1 to 48, we get the square number 49. If we add 1 to half of 48, namely, 24, we also get square number 25. The question that one might raise is to determine if there are other numbers that share this peculiar property. To offer entertainment and not frustration, we offer the next three larger numbers that have this unusual property. These numbers are as follows: 1,680; 57,120; and 1,940,448. We can see this in the following way:
  - $1,680 + 1 = 1,681 = 41^2$, and $1680 \div 2 = 840$, and $840 + 1 = 841 = 29^2$;
  - $57,120 + 1 = 57,121 = 239^2$, and $57,120 \div 2 = 28,560$, and $28,560 + 1 = 28,561 = 169^2$;
  - $1,940,448 + 1 = 1,940,449 = 1,393^2$, and $1,940,448 \div 2 = 970,224$, and $970,224 + 1 = 970,225 = 985^2$.
- The largest prime number less than 1 billion is 999,999,937.
- The number 85 can be expressed as the sum of two squares in two different ways, namely, $6^2 + 7^2 = 9^2 + 2^2$.
- The value of $48^4$ is equal to the product of all the factors 48. We can see this by considering the product of the factors of 48, that is, $1 \times 2 \times 3 \times 4 \times 6 \times 8 \times 12 \times 16 \times 24 = 5,308,416 = 48^4$.
- When you take the product of the digits of the number 39 and add it to the sum of these digits, you will end up with the number 39. Can you find other such numbers where this is true? (The answer is, *yes*, all two-digit numbers, whose units digit is a 9.)
- The number 6,534 is one and a-half times reversal. Symbolically, $6534 = \frac{3}{2} \times 4356$.
- Here is a cute number relationship. Ask your audience to find numbers to replace *a*, *b*, and *c*, so that $a^b \times c^d = a,bcd$. The only possible answer to this conundrum is $2^5 \times 9^2 = 2,592$.
- The number 3,435 is an unusual number, since it can be expressed as the sum of its digits each taken to the power equal to the digit.

Symbolically, that is $3{,}435 = 3^3 + 4^4 + 3^3 + 5^5$. Another number that has the same property is the number 438,579,088. Your audience may wish to verify that property.

- The number 175 can be expressed as the sum of the increasing powers of its digits: $175 = 1^1 + 7^2 + 5^3$. Others that have this property are as follows: $135 = 1^1 + 3^2 + 5^3$, $518 = 5^1 + 1^2 + 8^3$, and $598 = 5^1 + 9^2 + 8^3$.

- The number 1,634 can be expressed as the sum of each of its digits raised to the fourth power. Symbolically, that can be seen as $1634 = 1^4 + 6^4 + 3^4 + 4^4$. The only other 2 four-digit numbers where this is possible are 9474 and 8208.

- When the number 497 is doubled, one gets the reversal of the number 497 increased by 2. Symbolically, $497 \times 2 = 994$, and $497 + 2 = 499$, the reversal.

- We all know that there are 365 days in a non-leap year, however, the number 365 also has a very curious property, as shown here: $365 = 10^2 + 11^2 + 12^2 = 13^2 + 14^2$.

- A nice challenge for entertainment would be to find three-digit numbers which are prime, and where all arrangements of the three digits are also prime numbers. There are three such numbers, 113, 199, and 337. For example, suppose we take the first of these and rearrange the digits in all possible permutations: 113, 131, 311. Each of these three numbers is also a prime number.

- Look at the symmetry of these two equalities. Enjoy the symmetry!

$$1 + 5 + 6 = 2 + 3 + 7$$
$$1^2 + 5^2 + 6^2 = 2^2 + 3^2 + 7^2$$

Can you find other such symmetries?

- The number 31 has a curious property that can be best seen symbolically: $31 = 1 + 5 + 5^2 = 1 + 2 + 2^2 + 2^3 + 2^4$.

- It could be entertaining to notice that the number 132 provides the digits that will enable the audience to create different two digit numbers whose sum will be equal to 132. We can show that as follows: $12 + 13 + 21 + 23 + 31 + 32 = 132$.

- There are times when what appears to be a rather unusual phenomenon and can be quite entertaining because of its strange result (which, of course, can be very easily algebraically justified). It is essentially a rather simple camouflaged arithmetic result. If you select any number and add to it the next two consecutive numbers and then divide it by 3, you will get the middle number. For example, suppose we select the number 7, and then add the next two consecutive numbers: $7+8+9 = 24$ and then divide that by 3, which is the number of numbers we have, we will get the center number, $24 \div 3 = 8$. At this point, your audience may realize that you have calculated the average of the 3 numbers, but because it was presented in this rather unexpected fashion was well camouflaged.

Let's take that a step further and this time beginning with the numbers 7, 8, and 9 and then add the next two consecutive numbers. When we add these numbers: $7+8+9+10+11=45$, and then divide that by the number of numbers we just added we again get the middle number: $45 \div 5 = 9$.

Just to make sure that we are pushing on something that is consistent without actually doing the algebraic justification (which is being left to an ambitious reader), we will take the next two numbers and added to our previous sequence: $7+8+9+10+11+12+13 = 70$, and when divided by 7, yields the middle number 10.

This can go on for as long as you wish by each time beginning with any odd number of numbers and each time adding two numbers to the previous sum, dividing by the number of numbers being added to get you to the middle number of the sequence. By now a clever reader will recognize that the next two numbers added to the sequence will find the number 11 as the middle number and also achieved through this interesting calculation. Now that your audience may be enchanted by the way you presented this little entertaining aspect of arithmetic, you might mention to them that at this point what they have done is merely to find the average — which in this case is the middle number — of an odd number of consecutive numbers. This could have been found at any time by taking the sum and dividing by the number of numbers being added. Your audience should be delighted with this revelation.

- There are pairs of triplet-numbers whose products all use the same digits in different orders. These are:
  - $333 \times 777 = 258{,}741$,
  - $333 \times 444 = 147{,}852$, and
  - $666 \times 777 = 517{,}842$.

- By the way, the number 666 is often referred to as "the number of the beast" which could likely be from the Book of Revelation 13:15-18, which reads "And that no man might buy or sell, save he that had the mark, or the name of the beast, or the number of his name. Here is wisdom. Let him that hath understanding count the number of the beast: for it is the number of a man; and his number is Six hundred threescore and six." However, in mathematics, it has some interesting aspects. It is the sum of the squares of the first 7 prime numbers, that is, $666 = 2^3 + 3^2 + 5^2 + 7^2 + 11^2 + 13^2 + 17^2$. It is also the sum of the first 36 numbers: $1 + 2 + 3 + \cdots + 34 + 35 + 36 = 666$. Incidentally, when written in form, it uses all the Roman symbols prior to the M (1,000) in descending order: $666 =$ DCLXVI.

  The sum of the digits of 666 ($6 + 6 + 6 = 18$) is equal to the sum of the digits of its prime factors. Its prime factors are: $666 = 2 \cdot 3 \cdot 3 \cdot 37$, and the sum of the digits of the prime factors is: $2 + 3 + 3 + 3 + 7 = 6 + 6 + 6 = 18$.

  Now, if you really want to baffle your audience, tell them that the value of $\pi$ to 144 places is
  $\pi \approx 3.14159265358979323846264338327950288419716939937510582097494459230781640628620899862803482534211706798214808651328230664709384460955058223172 5359\ldots$
  The sum of these 144 digits is 666.

- You can also have fun by searching for curious arrangements of numbers. For example, here are five groups of three numbers, each of which shares the same product as the others, as well as the same sum as the other groups (see Figure 1.2).

| | Groups of Three Numbers | | | Sums | Products |
|---|---|---|---|---|---|
| Group 1 | 6 | 480 | 495 | 981 | 1,425,600 |
| Group 2 | 11 | 160 | 810 | 981 | 1,425,600 |
| Group 3 | 12 | 144 | 825 | 981 | 1,425,600 |
| Group 4 | 20 | 81 | 880 | 981 | 1,425,600 |
| Group 5 | 33 | 48 | 900 | 981 | 1,425,600 |

**Figure 1.2**

- Another entertaining aspect of arithmetic, which usually amazes the audience, is when numbers are placed differently and yet come up with analogous results. Admire the following:
  - $12^2 = 144$, and $21^2 = 441$. This is an oddity and should not be generalized.
  - $13^2 = 169$, and $31^2 = 961$. Similarly, this, too, should not be generalized.
  - $\sqrt{5\frac{5}{24}} = 5\sqrt{\frac{5}{24}}$. This can be justified as follow:

$$\sqrt{5\frac{5}{24}} = \sqrt{\frac{125}{24}} = \sqrt{\frac{25 \times 5}{24}} = 5\sqrt{\frac{5}{24}}$$

  - $\sqrt[3]{2\frac{2}{7}} = 2\sqrt[3]{\frac{2}{7}}$. This can also be justified as follows:

$$\sqrt[3]{2\frac{2}{7}} = \sqrt[3]{\frac{16}{7}} = \sqrt[3]{\frac{8 \times 2}{7}} = 2\sqrt[3]{\frac{2}{7}}$$

There are many other such awkward looking arrangements which tend to poke fun at arithmetic.

## The Numbers 9 and 11

The numbers 9 and 11 have very peculiar properties in the base 10 system, since they are on either side of the base number. For example,

$11^3 = 1331$, a palindromic number, and the number 9 also produces a very unusual pattern, since $9^3 = 729 = 1^3 + 6^3 + 8^3 = 3^6$.

Yet, the two numbers, 9 and 11, are tied together in other ways. Such as $\frac{1}{9} = 0.11111111111111111111...$ , and $\frac{1}{11} = 0.090909090909$ $0909...$. We know that the product of 9 and 11 is 99 and that also provides a rather unusual unit fraction: $\frac{1}{99} = 0.0101010101010101...$. While on the topic of 99, here is another curiosity that can be entertaining: $99^2 = 9801$, and if we split and add, $98 + 01 = 99$.

Taking the number 9 a step further considers the number $999 = 27 \times 37$. Now, by taking the reciprocals of these two numbers, another nice pattern and relationship evolves:

$$\frac{1}{27} = 0.037037037037037037...$$

and when we compare that pattern to

$$\frac{1}{37} = 0.027027027027027027...,$$

we notice a fascinating relationship between these two numbers that came from the factors of 999.

Not wanting to ignore 9's partner, the number 11, it also provides some curious patterns, such as

$$11 = 6^2 - 5^2$$
$$1111 = 56^2 - 65^2$$
$$111{,}111 = 556^2 - 445^2$$
$$11{,}111{,}111 = 5556^2 - 4445^2$$

Later, in this chapter, we will encounter other fascinating relationships and patterns involving numbers consisting of only 1's. Perhaps, we should further pursue these two important numbers, 11 and 9, in our base 10. This brief journey should be fascinating.

# Multiplying by 11

It is always easy to entertain folks by showing them clever shortcuts in doing arithmetic. These are the sorts of things that are generally enthusiastically transferred among friends to impress one another. Multiplying certain numbers can be done easily, if one is aware of the technique that can be used. One such was introduced in the introduction of this book, that is, multiplying a two-digit number by 11. Yet as we mentioned earlier, this technique can be broadened to multiplying larger numbers by 11. We will consider some now. First, let's review how to multiply a two-digit number by 11. All you need to do is to place the sum of the 2 digits between them to get product. For example, if we want to multiply 53 by 11, we would simply place the sum of $5 + 3 = 8$ between the 5 and the 3 to get the number 583, which is the appropriate product. Although it requires a bit more work, multiplying larger numbers can be done in the same way as we will show with a number such as 12,345, which we will multiply by 11. Here, we begin at the right-side digit and add every pair of digits going from right to left: $1[1 + 2][2 + 3][3 + 4][4 + 5]5 = 135{,}795$.

Suppose we now combine the skill we have garnered through this technique of multiplying by 11 and applying it to a number, which requires the more complicated version, where the sum of two adjacent digits exceeds 9. Remember, if the sum of two digits is greater than 9, then use the procedure described earlier, namely, place the units digit of this two-digit sum appropriately and carry the tens digit to the next place. To best illustrate how this technique would be used with a larger number, we will do one of these more complicated versions here. Let us consider multiplying 56,789 by 11. This may be a little bit tedious, and perhaps somewhat less realistic for common use, but we will show it here merely to demonstrate the extension of this multiplication technique, so that the audience gets a fuller picture of multiplying by 11 mentally. Follow along, as we do this step-by-step.

| | |
|---|---|
| 5[5 + 6][6 + 7][7 + 8][8 + 9]9 | Add each pair of digits between the end digits |
| 5[5 + 6][6 + 7][7 + 8][17]9 | Add 8 + 9 = 17 |
| 5[5 + 6][6 + 7][7 + 8 + 1][7]9 | Carry the 1 (from the 17) to the next sum |
| 5[5 + 6][6 + 7][16][7]9 | Add 7 + 8 + 1 = 16 |
| 5[5 + 6][6 + 7 + 1][6][7]9 | Carry the 1 (from the 16) to the next sum |
| 5[5 + 6][14][6][7]9 | Add 6 + 7 + 1 = 14 |
| 5[5 + 6 + 1][4][6][7]9 | Carry the 1 (from the 14) to the next sum |
| 5[12][4][6][7]9 | Add 5 + 6 + 1 = 12 |
| 5 + 1[2][4][6][7]9 | Add 1 + 5 = 6 to get the answer: 624,679 |

We need to keep in mind that for anything that we present to be entertaining it is important that the mode of presentation is done with personal enthusiasm, something that can be very contagious — in a favorable way. This is not to be necessarily seen as an instructional session, rather an entertaining one, so it should be done in a lighthearted fashion and only extended to the multi-digit numbers if the audience requests it. Remember, entertainment is the key factor in this presentation.

## Divisibility by 11

While we have entertained the audience with a clever way to multiply a number by 11, it may be appropriate to ask the audience if they would like to see a clever way of determining whether a given number is divisible by 11. This can only be entertaining, if it is presented in such a way as a direct follow-up to the previous surprising technique and then offered in a lighthearted fashion. You might mention that the reason that the number 11 provides is entertaining aspects is because it is 1 greater than the base of our number system, 10. Clearly, only at the oddest times will the need come up to determine if a number is divisible by 11. If you have a calculator at hand, the problem is easily solved. But that is not always the case. Besides, there is such a clever "rule" for testing for divisibility by 11 that it is worth knowing just for its charm, and as a result can be presented in a rather entertaining fashion.

The technique is quite simple:

*If the difference of the sums of the alternate digits is divisible by 11, then the original number is also divisible by 11.*

This sounds more complicated than it is. Let us take this process a piece at a time. The sum of the alternate digits means that you begin at one end of the number taking the first, third, fifth, etc. digits and add them. Then take the sum of the remaining (even placed) digits. Subtract the two sums and inspect for divisibility by 11. If the resulting number is divisible by 11, then the original number was divisible by 11. And the reverse is also true. That is, if the number reached by subtracting the two sums is *not* divisible by 11, then the original number was also *not* divisible by 11. Since this is being presented as entertainment, it might be wise to demonstrate this technique through an example.

Suppose we test the number 768,614 for divisibility by 11. Sums of the alternate digits are $7 + 8 + 1 = 16$, and $6 + 6 + 4 = 16$. The difference of these two sums, $16 - 16 = 0$, which is divisible by 11. (Remember $\frac{0}{11} = 0$.) Therefore, we can conclude that 768,614 is divisible by 11. Of course, depending on the audience, full understanding this technique is essential for it to be entertaining. Therefore, having the audience consider another example might be worthwhile. To determine if 918,082 is divisible by 11, we need to find the sums of the alternate digits: $9 + 8 + 8 = 25$, and $1 + 0 + 2 = 3$. Their difference is $25 - 3 = 22$, which is divisible by 11, and so the number 918,082 is divisible by 11. The audience might like to practice this technique so that they can share this curiosity with others.

Sometimes entertainment leads to a need to know what is actually happening to make this work as it does. In this case, the audience might request a justification for this unusual process. This may not be as entertaining as the original-intended demonstration; however, it might serve well to satisfy curious audience.

To justify this procedure algebraically, we will consider the number to be tested for divisibility by 11 as *ab,cde*, whose value can be expressed as follows: (Note: 11*M* refers to a multiple of 11.)

$$N = 10^4 a + 10^3 b + 10^2 c + 10 d + e$$
$$= (11-1)^4 a + (11-1)^3 b + (11-1)^2 c + (11-1)d + e$$
$$= [11M + (-1)^4]a + [11M + (-1)^3]b + [11M + (-1)^2]c + [11 + (-1)]d + e$$
$$= 11M[a+b+c+d] + a - b + c - d + e$$

which implies that since the first part of this value of *N*, namely, $11M[a+b+c+d]$, is already a multiple of 11, the divisibility by 11 of *N* depends on the divisibility of the remaining part, $a - b + c - d + e = (a + c + e) - (b + d)$, which is actually the difference of the sums of the alternate digits. Remember, the justification just presented should only be shown on request, otherwise it may detract from the entertaining aspect of this technique.

As a further challenge, it might be nice to try to find the number, which is 11 times a sum of its digits. (It may be helpful to look back to the technique for multiplying by 11.) It might take a little bit of effort, but a clever audience should see that the sought-after number as 198, since 198 = $11 \times (1 + 9 + 8)$. Again some hidden gems with which to entertain!

## The Number 9

In the base 10, we showed some curious entertainments with the number 11, which was 1 greater than the base 10 and now we can also do some entertaining with a number 9, which is 1 less than the base 10. There are moments in everyday life situations where it can be useful to know, if a number can be divisible by 3 or by 9, especially if it can be done instantly "in your head," such as in a restaurant, if a bill needs to be split evenly among three people, it would be nice to know if this can be done equally. To entertain your audience, it would be wise to think of a situation appropriate for that audience so that they would see the forthcoming technique entertaining and useful. For example, assume you are in a restaurant and received a check of $71.23, and you want to add a tip, but you would want the end result to be such that it could be split into three equal parts. Wouldn't it be nice if there were some clever arithmetic technique for doing this mentally and instantly. Well, here comes mathematics to the rescue. We are going to provide you with a simple technique to determine if a number is divisible by 3 (and as an extra bonus, also divisible if it is by 9).

The entertaining part of this illustration is the simplicity with which the result can be obtained. The technique is as follows:

*If the sum of the digits of a number is divisible by 3 (or 9), then the original number is divisible by 3 (or 9).*

As before, perhaps an example would best firm up an understanding of this technique. Consider the number 296,357. Let's test it for divisibility by 3 (or 9). The sum of the digits is $2 + 9 + 6 + 3 + 5 + 7 = 32$, which is not divisible by 3 or 9. Therefore, the original number is neither divisible by 3 nor by 9. On the other hand, if we want to determine if the number 548,622 is divisible by 3 or by 9, we once again take the sum of the digits which in this case is 27, which is divisible by 3 and by 9 and therefore, the number 548,622 is divisible by 3 and by 9.

Let's now assume a group of three people is given a restaurant check of $71.23, and would like to give an approximate 20% tip on this check. They decide to add $14 to the bill, which would make the total $85.23. You now tell your audience that they would just like to know if this last number $85.23 can be divided equally among the three people. The procedure would have them add the digits to get: $8 + 5 + 2 + 3 = 18$, which is divisible by three, and therefore, would allow them to equally divide the check among the three customers. If the number reached from the sum of the digits is not easily identifiable as a multiple of 3, then continue to add the digits of that resulting number until you reach a number, which you can visually recognize as a multiple 3. In this case, it should be noted that the final result, 18, is also divisible by 9, which implies that the original number $85.23 was divisible by 9 as well.

Once again, the audience might want you to explain why this trick works as it does. To provide you, the entertainer, with the proper information, we offer a brief discussion as to why this rule works as it does. Consider the number *ab,cde*, whose value can be expressed as (where 9*M* refers to a multiple of 9)

$$N = 10^4 a + 10^3 b + 10^2 c + 10d + e$$
$$= (9+1)^4 a + (9+1)^3 b + (9+1)^2 c + (9+1)d + e$$
$$= [9M + (1)^4]a + [9M + (1)^3]b + [9M + (1)^2]c + [9 + (1)]d + e$$
$$= 9M[a + b + c + d] + a + b + c + d + e$$

which implies that divisibility by 9 of the number *N* depends on the divisibility of $a + b + c + d + e$, which is the sum of the digits, since the first part of the equation, $9M[a + b + c + d]$ it is already a multiple of 9.

Typically, an audience would be quite enthralled by the simplicity and usefulness of this procedure. Therefore, they may want to try to apply it to a few numbers. Here are two examples that you may wish to use. Is the number 745,785 divisible by 3 or 9? The sum of the digits is $7 + 4 + 5 + 7 + 8 + 5 = 36$, which is divisible by 9 (and then, of course, divisible by 3 as well), so the number 745,785 is divisible by 3 and by 9.

It is also possible to get a result as with the following example: Is the number 72,879 divisible by 3 or 9? The sum of the digits is $7 + 2 + 8 + 7 + 9 = 33$, which is divisible by 3 but not by 9; therefore, the number 72,879 is divisible by 3 and not by 9. Oftentimes, such useful little schemes are also quite entertaining.

## Find the Missing Number

Working with all 10 digits is sometimes amusing. Have your audience write down 2 numbers using all 10 digits exactly once. While they do that, we will do it here as well. We are selecting the following 2 numbers 87,503 and 2,146. Notice we used all 10 digits to form these 2 numbers.

Keep in mind, you should not see the 2 numbers they created. Now have them add the 2 numbers, as we will do as well, to get our sum of 89,649. (Remember, you still do not know what numbers they have selected nor what they sum is.) Now ask them to delete any digit from their resulting sum, as we will do here, by deleting the number 6. Next, ask your audience to find the sum of the digits of the remaining number from the sum that is now missing one digit. The sum of our digits is $8 + 9 + 4 + 9 = 30$. All we need to do now is to find the number greater than 30 which is divisible by 9. That number is 36, which is 6 greater than 30, and therefore, the number deleted was the 6. Therefore, when they reveal their final digit sum, you can determine how far the next multiple of 9 is, and that will give you the number they have deleted.

In case you're not convinced that this will work consistently, you may wish to test this technique again by supposing we had deleted the number 8 from our earlier sum, we would then be left with $9 + 6 + 4 + 9 = 28$, which requires an 8 to reach the next multiple of 9 which is 36. And so we have discovered the deleted number 8.

This is very closely related to the above discussion of any number which is a multiple of 9 has the sum of its digits also a multiple of 9, and since the sum of the initial digits used is $0 + 1 + 2 + 3 + 4 + 5 + 6 + 7 + 8 + 9 = 45$, which is a multiple of 9, the sum of any 2 numbers using all of these 10 digits will also be a multiple of 9. This should clear up your understanding of this technique.

## A Peculiarity of the Units Digit 9

There is an unusual peculiarity in the base 10 number system, where all two-digit numbers having a units digit of 9 will be equal to the sum and the product of their digits. This is extremely easy to explain to your audience and will certainly get a "geewhiz" reaction.

Take, for example, the number 29, and look what happens when we add the sum and the product of its digits:

$$\text{For 29: } (2 + 9) + (2 \times 9) = 11 + 18 = \mathbf{29}$$

To show this works beyond this one number, here are two more examples to justify this peculiar relationship:

$$\text{For 49: } (4 + 9) + (4 \times 9) = 13 + 36 = \mathbf{49}$$
$$\text{For 79: } (7 + 9) + (7 \times 9) = 16 + 63 = \mathbf{79}$$

An ambitious audience might want to check if other two-digit numbers ending in nine also follow this pattern. With such simple peculiarities, folks like to pass that along to their friends to impress them.

## More Fun with the Number 9

The number 9 provides us with a host of other entertainments, one of which we encountered in the introduction to this book. Some of the entertainments results from the divisibility technique we discussed earlier. For example, you might want to impress your audience by generating the number 9 from various numbers that they will provide you with. Begin by selecting someone from your audience to choose any number and add it to his or her age. Then, have him or her add to this result the last two digits of his or her telephone number. At this point your audience is probably bewildered as to where you are headed. Next ask this person to multiply this resulting number by 18. Then have them obtain the sum of the digits of this last number. Once again, the sum of the digits of this previously resulting sum should be found and then continue to take the digit sums of each resulting number until a single digit is reached. You will impress your friends by telling them that this last digit obtained is a 9. A clever audience might already realize why this has happened, especially after they have been exposed to the previously presented characteristics of the number 9.

To better understand this technique, let's try this with the initial selection of the number 39 and add our age (37) to it to get 76. We then add the last two digits of our telephone number (31) to get 107 and then multiply this number by 18 to get 1926. The digit sum is $1 + 9 + 2 + 6 = 18$, whose digit sum is $1 + 8 = 9$. The reason that this works is because when we multiplied by 18, we made sure that the final result would be a multiple of 9, which eventually always yields a digit sum of 9. So there we have another simple entertaining activity stemming from our previous experience with the number 9 that will surely impress your audience.

To further exploit the number 9 for entertainment purposes, we offer one that can be used right across the dinner table, since it can be done mentally and very easily at that. It goes this way: Have your audience select a number from 1 to 10, and have them keep it secret. Then ask them to multiply their secret number by 9. Then ask them to add the digits of the number they arrived at and subtract 5 from that number. Still without divulging any information, have them select the letter of the alphabet that corresponds with the number they arrived at. In other words, $A = 1$, $B = 2$, $C = 3$, etc. Have them select a country in Europe, whose first letter is the same as the one that they have identified? Then ask them to select an animal whose name begins with the last letter of the European country they identified. You will be able to tell them that the animal that they have identified is a kangaroo. They will surely be amazed at your talent.

Here is the reason for your hidden "talent." When they added the digits of the number that they found by multiplying the original number by 9, the result will have to have been a 9, as we have already discussed earlier. Subtracting 5 from the 9 yields 4. The fourth letter of the alphabet is a $D$. The only country in Europe whose name begins with a $D$ is Denmark. The only *popular* animal whose name begins with the last letter of the word Denmark, a $K$, is the kangaroo. And so, you have used mathematics to entertain your audience. (N.B. in rare instances someone in the audience might have come up with the animal Kakapo, or King Crab, Koala, Kingfisher, or other such exotic animals, but that is hardly likely.)

## More Oddities of the Number 9

There are times when there is little space for entertainment, so it is sometimes cute to just show off a few peculiarities of a number. The number 9 lends itself very well to such oddities as you can see below. Here are some such oddities:

- The only square number that is the sum of 2 consecutive cubes, namely $9 = 1^3 + 2^3$.
- Number 9 can be expressed as a sum of 3 consecutive factorials, such as $9 = 1! + 2! + 3!$
- The number 9 is the smallest Kaprekar number after the number 1, where $9^2 = 81$ and $8 + 1 = 9$. (See page 80).
- How large can a number be using only three 9s? Using only single digits, the largest number we can make is $9^{9^9}$ which is actually $9^{387,420,489}$, which results in a number that is 369,693,100 digits long. Imagine how long a piece of paper must be to accommodate this number. As a matter of fact, the famous German mathematician Carl Friedrich Gauss considered the following number an "unmeasurable infinity:" $9^{9^{9^9}}$

While still on the number 9 there is an arithmetic check that can be not only entertaining but also useful when a calculator is not at hand. We often check addition by adding in a different direction and hoping to get the same result. However, in the next section, we offer a curious way of checking addition called "casting out nines."

## Casting out Nines

We have much for which we can thank Leonardo of Pisa — more popularly known as Fibonacci — since much of what we have and do in mathematics today can be traced back to his brilliance. It is nice to share a bit of history of common usage to entertain the general audience. For example, the numeral system that we use today, namely, 1, 2, 3, 4, 5, 6, 7, 8, 9, and 0 was first introduced to the European world, in Fibonacci's book *Liber Abaci*, which was first published in the year

1202. The first sentences in Chapter 1 of this book read as follows: "The nine Indian figures are 9, 8, 7, 6, 5, 4, 3, 2, 1. With these nine figures, and the sign 0, which the Arabs call Zephir, any number whatsoever can be written,..." It is quite likely that Fibonacci came across these numbers when he spent time in his youth accompanying his father, who was a public official in the Bugia (a port city in eastern Algeria) customs house established for the Pisan merchants. There Fibonacci was taught mathematics using these symbols and so he incorporated them into his book, which was republished in 1228. Yet it is believed, that it wasn't until 50 years later that these numerals became more universally used in Europe. We also credit Fibonacci as the first to use the fraction bar to represent a fraction such as $\frac{1}{2}$.

In Chapter 3 of his book *Liber Abaci*, Fibonacci introduces a procedure for testing the correctness of an addition computation, which he called "casting out nines." As you have enlightened your audience to this point with some unexpected information about our common number usage, you can extend the entertainment by demonstrating how this procedure of casting out nines actually works. All we need to do is to find the sum of the digits of each of the addends and then take the sum of the sums and compare it to the sum of the digits of the computational sum of numbers. If they are the same, then the answer is correct.

We can show this with the following addition computation:

13579 — digit sum: 25 — digit sum: 7

+86327 — digit sum: 26 — digit sum: 8

**sum of digit sums**: $7 + 8 = 15$ — digit sum: 6

99906 — digit sum: 33 — digit sum: 6

Since the sum of the digit sums of the two addends eventually is 6, and it matches the digit sum of the addition calculation sum, also 6, the addition could be correct.

It should be noted that Fibonacci is far more famous today for the Fibonacci numbers which appear in Chapter 12 of *Liber Abaci*, and which lend themselves to a plethora of entertaining activities which we will address later.

## Multiplying *Our* Numbers by Multiples of 9

The number 9 and its multiples produce a rather unexpected pattern of results when multiplied by a number formed with our 9 numerals in descending order. Once again, the calculator may be required to make this event run smoothly. It requires quite a bit of multiplication before the beauty of the patterns emerges.

Multiplying 987654321 by multiples of 9

$$987654321 \times 9 = 8888888889$$
$$987654321 \times 18 = 17777777778$$
$$987654321 \times 27 = 26666666667$$
$$987654321 \times 36 = 35555555556$$
$$987654321 \times 45 = 44444444445$$
$$987654321 \times 54 = 53333333334$$
$$987654321 \times 63 = 62222222223$$
$$987654321 \times 72 = 71111111112$$
$$987654321 \times 81 = 80000000001$$
$$987654321 \times 90 = 88888888890$$
$$987654321 \times 99 = 97777777779$$
$$987654321 \times 108 = 106666666668$$
$$987654321 \times 117 = 115555555557$$
$$987654321 \times 126 = 124444444446$$
$$987654321 \times 135 = 133333333335$$
$$987654321 \times 144 = 142222222224$$
$$987654321 \times 153 = 151111111113$$
$$987654321 \times 162 = 160000000002$$
$$987654321 \times 171 = 168888888891$$
$$987654321 \times 180 = 177777777780$$
$$987654321 \times 189 = 186666666669$$

$$987654321 \times 198 = 195555555558$$
$$987654321 \times 207 = 204444444447$$
$$987654321 \times 216 = 213333333336$$
$$987654321 \times 225 = 222222222225$$
$$987654321 \times 234 = 231111111114$$
$$987654321 \times 243 = 240000000003$$
$$987654321 \times 252 = 248888888892$$
$$987654321 \times 261 = 257777777781$$
$$987654321 \times 270 = 266666666670$$
$$987654321 \times 279 = 275555555559$$
$$987654321 \times 288 = 284444444448$$
$$987654321 \times 297 = 293333333337$$
$$987654321 \times 306 = 302222222226$$

An ambitious reader may wish to continue to watch this pattern grow impressively. Patterns such as these, which are completely unexpected serve well to exploit the hidden beauty of mathematics.

Yet, we can take this a step further by reversing the order of the consecutive digits without the number 8 and multiplying by multiples of 9 and discover how another surprising pattern of numbers emerges, as shown in Figure 1.3.

| 12345679 | × | 9 | = | 111,111,111 |
|---|---|---|---|---|
| 12345679 | × | 18 | = | 222,222,222 |
| 12345679 | × | 27 | = | 333,333,333 |
| 12345679 | × | 36 | = | 444,444,444 |
| 12345679 | × | 45 | = | 555,555,555 |
| 12345679 | × | 54 | = | 666,666,666 |
| 12345679 | × | 63 | = | 777,777,777 |
| 12345679 | × | 72 | = | 888,888,888 |
| 12345679 | × | 81 | = | 999,999,999 |

**Figure 1.3**

# More Strange Multiples of 9

Multiplication with large numbers has become trivial using the calculator. However, some numbers have unusual properties and remain so after multiplication, especially when multiplied by the number 9, as we have seen in the previous example. It might be challenging to find numbers that are composed of 8 different numerals and when multiplied by 9 results in numbers consisting of 9 different numerals.

This can be an interesting challenge for your audience. However, generously we offer several such examples here:

$$58132764 \times 9 = 523194876$$

$$76125483 \times 9 = 685129347$$

$$72645831 \times 9 = 653812479$$

$$81274365 \times 9 = 731469285$$

What is particularly notable here, is that the numeral 9 is missing from each number being multiplied by 9, and yet appears in each of the end products. Had we multiplied these original numbers by 18, we would be able to get numbers with 10 different digits; however, this time the 0 would be included, which was excluded from the above products. Here are the products, each consisting of 10 different digits:

$$58132764 \times 18 = 1046389752$$

$$76125483 \times 18 = 1370258694$$

$$72645831 \times 18 = 1307624958$$

$$81274365 \times 18 = 1462938570$$

As neat as this appears, it still would be a challenge for your audience to find other such eight-digit numbers, that when multiplied by 9 result in products, which have 9 different digits, and when multiplied by 18 result in products, which have 10 different digits. An interesting exercise can be entertaining if presented properly.

## More Fun with 9s

There are times when you can amuse an audience with some unusual patterns in mathematics. The 9s lend themselves nicely to such a situation as you can see from Figure 1.4.

| | | | | |
|---:|:---:|---:|:---:|---:|
| 9 | × | 9 | = | 81 |
| 99 | × | 99 | = | 9,801 |
| 999 | × | 999 | = | 998,001 |
| 9,999 | × | 9,999 | = | 99,980,001 |
| 99,999 | × | 99,999 | = | 9,999,800,001 |
| 999,999 | × | 999,999 | = | 999,998,000,001 |
| 9,999,999 | × | 9,999,999 | = | 99,999,980,000,001 |

**Figure 1.4**

In Figure 1.5 you can demonstrate how a pattern of numbers when multiplied by 9 and added to consecutive numbers result in a rather unexpected pattern of 8s.

| | | | | | | |
|---:|:---:|:---:|:---:|:---:|:---:|:---|
| 0 | × | 9 | + | 8 | = | 8 |
| 9 | × | 9 | + | 7 | = | 88 |
| 98 | × | 9 | + | 6 | = | 888 |
| 987 | × | 9 | + | 5 | = | 8,888 |
| 9,876 | × | 9 | + | 4 | = | 88,888 |
| 98,765 | × | 9 | + | 3 | = | 888,888 |
| 987,654 | × | 9 | + | 2 | = | 8,888,888 |
| 9,876,543 | × | 9 | + | 1 | = | 88,888,888 |
| 98,765,432 | × | 9 | + | 0 | = | 888,888,888 |

**Figure 1.5**

This time we take a factor/multiples of 9 and multiply them by the number 37,037, as shown in Figure 1.6, to once again get some surprising results that clearly should impress the audience.

$$37037 \times 3 = 111111$$
$$37037 \times 9 = 333333$$
$$37037 \times 6 = 222222$$
$$37037 \times 18 = 666666$$
$$37037 \times 12 = 444444$$
$$37037 \times 15 = 555555$$

**Figure 1.6**

## A Number Generated by 9s

Another number that yields some symmetric results is obtained when we divide 999,999 by 7, which is 142857. This time we will multiply the number 142857 sequentially by 1, 3, 2, 6, 4, and 5 to get the following products:

$$
\begin{aligned}
142857 \times 1 &= 142857 \\
142857 \times 3 &= 428571 \\
142857 \times 2 &= 285714 \\
142857 \times 6 &= 857142 \\
142857 \times 4 &= 571428 \\
142857 \times 5 &= 714285
\end{aligned}
$$

Careful inspection of these products will show that the first and last digits of each of the products to be 1, 4, 2, 8, 5, 7 and the diagonal from upper right to lower left consists of all 7s and similar patterns as previously shown are embedded in this group of products. Such unusual number patterns will always be well received by your audience.

## Square Number Patterns

We can begin with a very unexpected pattern that seems to surprisingly unveil square numbers. This will not only impress your audience, but also provide food for further thought. Begin, by having them write the sequence of natural numbers in groups as shown below:

$$
\begin{aligned}
&1 \\
&2,3 \\
&4,5,6 \\
&7,8,9,10 \\
&11,12,13,14,15 \\
&16,17,18,19,20,21 \\
&22,23,24,25,26,27,28
\end{aligned}
$$

Then have them cancel out every second group as we have done below, so that they are left with:

1

~~2,3~~

4,5,6

~~7,8,9,10~~

11,12,13,14,15

~~16,17,18,19,20,21~~

22,23,24,25,26,27,28

- If we take the sum of the first two remaining groups, we obtain

$$(1) + (4 + 5 + 6) = 16 = 4^2, \text{ or can be written as } 2^4$$

- If we take the sum of the first three remaining groups, we obtain

$$(1) + (4 + 5 + 6) + (11 + 12 + 13 + 14 + 15) = 81 = 9^2,$$
$$\text{or can be written as } 3^4$$

- If we take the sum of the first four remaining groups, we obtain

$$(1) + (4 + 5 + 6) + (11 + 12 + 13 + 14 + 15) + (22 + 23 + 24 + 25 + 26 + 27 + 28) = 256 = 16^2, \text{ or can be written as } 4^4$$

Your audience should notice that the relationship of the three results that we obtained formed a pattern. Ambitious participants in your audience will want to know if this pattern continues and you can assure them that it does. Another one of mathematics' hidden beauties.

## The Fibonacci Numbers

The Fibonacci numbers are perhaps a source of endless entertainment this is evidenced by a journal which was initiated in 1963 entitled The Fibonacci Quarterly, which to the present day generates a wide variety of articles related to these numbers. There are also entire books written about these famous numbers (for example, *The Fabulous Fibonacci Numbers* A. S. Posamentier and I. Lehmann, Prometheus Books, 2007). The Fibonacci numbers, 1, 1, 2, 3, 5, 8, 13, 21, 34, 55, 89, 144, ... were originally generated in a problem of the regeneration of rabbits in Chapter 12 of Leonardo of Pisa's (also known as Fibonacci) book *Liber Abaci,* first published in 1202.

However, the numbers lend themselves two observations in nature, such as the three spirals on a pineapple or the spirals on a pinecone, which are all Fibonacci numbers. The Fibonacci numbers can be useful in converting kilometers to miles and the reverse by simply going up one number or down one number in the sequence. For example, a good estimate of the number of kilometers covered when traveling 34 miles would be the next number up in the sequence, namely, 55 km. This can also work in reverse if one is traveling 34 km that would be the equivalent roughly of 21 miles.

There are also endless relationships that can be discovered among the Fibonacci numbers. Suppose we take the squares numbers of the Fibonacci numbers, such as, 1, 4, 9, 25, 64, 169, 441, ... . Now if we add the consecutive squares $1+4$, $4+9$, $9+25$, $25+64$, $64+169$, $169+441$, we get 5, 13, 34, 89, 233, 610, ..., which are the odd-positioned Fibonacci numbers beginning with the fifth term. If this is an amazing enough, you can further enthrall your audience by telling them to subtract the consecutive numbers in the last sequence, so as to get: 8, 21, 55, 144, 377,....

The cubes also play a helpful role in finding a relationship among the Fibonacci numbers. If we referred to the general Fibonacci number as $F_n$, we can generate Fibonacci numbers using cubes in the following way: $F_n^3 + F_{n-1}^3 - F_{n+1}^3 = F_{3n}$. We can apply this to a randomly selected sequence of Fibonacci Numbers, say the $F_5, F_6, F_7$ so that: $F_6^3 + F_7^3 - F_5^3 = F_{18}$, which yields: $512 + 2197 - 125 = 2584$. There other enchantments among the Fibonacci numbers with higher powers as well as with other relationships of these powers that we have shown. We encourage the reader to pursue these for further entertainment.

These are just little tidbits to whet your audience's appetite to delve into the Fibonacci numbers in greater depth. The entertainments there are probably limitless and go from very simple arithmetic relationships to geometric relationships as seen on the Golden rectangle which is generated by the Golden ratio and which can be determined by taking the ratio of consecutive Fibonacci numbers. The larger the Fibonacci numbers in the sequence are the close of the ratio approaches the Golden ratio. We have not even scratched the surface of what the Fibonacci numbers can provide entertainment, so go for it!

## Sum of Squares Equals Sum of More Squares

There are times when we like to challenge our audience. Here is one that sounds very simple and yet requires a little bit of thought. Have your audience take the sum of any three squares and multiply it by 3. They now need to find four squares that will have the same sum.

For example, $3(2^2 + 3^2 + 4^2) = 87 = 9^2 + 2^2 + 1^2 + 1^2$. Or perhaps as another example: $3(2^2 + 3^2 + 3^2) = 66 = 5^2 + 4^2 + 4^2 + 3^2$. This can be sometimes rather frustrating and yet also delightfully challenging, which is then considered entertaining. For the experts who might want to see a justification so that they don't feel that we left them with an unsolvable situation, we provide a simple algebraic proof:

$$3(a^2 + b^2 + c^2) = (a+b+c)^2 + (b^2 - 2bc = c^2) + (c^2 - 2ca + a^2)$$
$$+ (a^2 - 2ab + b^2)$$
$$= (a+b+c)^2 + (b-c)^2 + (c-a)^2 + (a-b)^2.$$

Using this relationship will also provide possible solutions.

## Getting Accustomed to Powers

A nice challenge which leaves the audience entertained and enhanced, is to ask them to provide 2 numbers that are not powers of 10 and what is the doctor say about your heart will yield a product of 1 billion? At first glance this will overwhelm the audience, however, they should be of quick to realize that 1 billion can be represented as $10^9$. We know that $10^9 = [(2)(5)]^9 = (2^9)(5^9) = (512)(1,953,125)$ and there you have it! This can be extended to other powers of 10 in a similar fashion, and will allow the audience to entertain others, such as $10^{18} = (2^{18})(5^{18}) = (262,144)(3,814,697,265,625)$.

(Note: the powers of 5 will always end in 5, 25, 125, or 625).

## Sums of Powers

For those who saw the movie "The Man Who Knew Infinity" they will recall the last scene where the famous Indian mathematician Srinivasa Ramanujan (1887–1920) instantly cites from his hospital bed that the number 1729 is the smallest number that can be expressed as a sum of two cubes in two different ways. That is, $1729 = 12^3 + 1^3 = 10^3 + 9^3$. By the way, the number 1729 is a number which is divisible by the sum of its digits. That is, $\frac{1729}{1+7+2+9} = \frac{1729}{19} = 91$. From this we also have another curiosity: $1729 = 19 \times 91$.

Building on the previous curiosity, the number 6578 is the smallest number that can be expressed as a sum of 3 fourth powers in 2 different ways. That is, $6578 = 1^4 + 2^4 + 9^4 = 3^4 + 7^4 + 8^4$. As a follow-up, the audience may search for two-digit numbers that can be expressed as a sum of squares in two different ways. One such number is $65 = 8^2 + 1^2 = 7^2 + 4^2$. By the way, the number 65 can also be expressed as sum of two cubes: $65 = 4^3 + 1^3$. By now you should be able to see how we can constantly seek such relationships among our numbers. Just searching for such number patterns can be quite amusing and upon success rewarding at the same time.

While we are fixed on the sum of squares, we can really find a curious arrangement as follows. If we multiply the sum of 2 different squares by the sum of 2 other different squares, the result would be the sum of 2 squares into different ways. We can show this symbolically in the following fashion:

$$(a^2 + b^2) \cdot (c^2 + d^2) = (ac + bd)^2 + (ad - bc)^2 \quad \text{or} \quad (a^2 + b^2) \cdot (c^2 + d^2)$$
$$= (ac - bd)^2 + (ad + bc)^2$$

Let's see how that works with the numbers $a = 2$, $b = 5$, $c = 3$, and $d = 6$. So that

$$(2^2 + 5^2) \cdot (3^2 + 6^2) = 29 \cdot 45 = 1305$$

We then can set up the following:

$$(2^2 + 5^2) \cdot (3^2 + 6^2) = (2 \cdot 3 + 5 \cdot 6)^2 + (2 \cdot 6 - 3 \cdot 5)^2 = 36^2 + (-3)^2$$
$$= 1296 + 9 = 1305$$

and alternatively:

$$(2^2 + 5^2) \cdot (3^2 + 6^2) = (2 \cdot 3 - 5 \cdot 6)^2 + (2 \cdot 6 + 3 \cdot 5)^2 = 24^2 + 27^2$$
$$= 576 + 729 = 1305$$

Without showing your audience the technique for getting pairs of squares, you might like to have them try it and see if they can come up with another set of two sums of squares. Before too much frustration sets in, you might want to show them how this can be done with the algebraic relationship shown above.

## A Surprising Pattern of Odd Numbers

Odd numbers can be presented in a way that they generate cubic numbers. This can be quite surprising as well as enchanting for the unsuspecting audience. The way it is presented can make the difference in its appreciation. So, it is left up to the reader to come up with a clever way of generating these results.

| | |
|---|---|
| $1 =$ | $1 = 1^3$ |
| $3 + 5 =$ | $8 = 2^3$ |
| $7 + 9 + 11 =$ | $27 = 3^3$ |
| $13 + 15 + 17 + 19 =$ | $64 = 4^3$ |
| $21 + 23 + 25 + 27 + 29 =$ | $125 = 5^3$ |
| $31 + 33 + 35 + 37 + 39 + 41 =$ | $216 = 6^3$ |

You might want to ask your audience what the sum of the tenth line will be. They should have been able to determine at this point that the tenth line would have a sum equal to $10^3 = 10,000$. A typical reaction from an audience is amazement how such a simple relationship can generate cubes.

## Some More Cute Number Patterns

It is not very difficult to notice that one can express a number as the sum of three other numbers. However, with the number 118, we can also express it as the sum of four arrangements of three numbers, and the amazing thing is that the product of each of these sets of three numbers is the same for all four groups of three, namely, 37,800. Take a look here:

$$15 + 40 + 63 = 118, \text{ and } 15 \times 40 \times 63 = 37,800$$
$$14 + 50 + 54 = 118, \text{ and } 14 \times 50 \times 54 = 37,800$$
$$21 + 25 + 72 = 118, \text{ and } 21 \times 25 \times 72 = 37,800$$
$$18 + 30 + 70 + 118, \text{ and } 18 \times 30 \times 70 = 37,800$$

More amazing: this number 118 is the smallest number for which this can be done. You might want to challenge your friends to come up with some other such arrangements for other numbers.

Here is another nice number relationship with which you can entertain others. Look at the symmetry:

$$13^3 - 3^7 = 2197 - 2187 = 13 - 3$$
$$5^3 - 2^7 = 125 - 128 = -(5 - 2)$$

Let them appreciate the symmetry, where a pair of powers can achieve this pattern. Please do not ask your audience to find another such pair of numbers, because no other such pair has yet been found!

There are many cute number relationships with which you can entertain your audience. We will present a few of them here. Notice that in each case the exponents are consecutive:

$$43 = 4^2 + 3^3$$
$$63 = 6^2 + 3^3$$

$$135 = 1^1 + 3^2 + 5^3$$
$$175 = 1^1 + 7^2 + 5^3$$
$$518 = 5^1 + 1^2 + 8^3$$
$$598 = 5^1 + 9^2 + 8^3$$
$$1306 = 1^1 + 3^2 + 0^3 + 6^4$$
$$1676 = 1^1 + 6^2 + 7^3 + 6^4$$
$$2427 = 2^1 + 4^2 + 2^3 + 7^4$$

And here is one where the exponents match the base:

$$3435 = 3^3 + 4^4 + 3^3 + 5^5$$

You might ask your audience to see if there are other such relationships, or even some such as
$244 = 1^3 + 3^3 + 6^3$ and $136 = 2^3 + 4^3 + 4^3$. Notice the relationship here!

In the event that you might want to keep this to one exponent number, you can do this with the following numbers:

$$153 = 1^3 + 5^3 + 3^3$$
$$370 = 3^3 + 7^3 + 0^3$$
$$371 = 3^3 + 7^3 + 1^3$$
$$407 = 4^3 + 0^3 + 7^3$$

We can take this even a step further by considering the four-digit numbers, such as the following:

$$1634 = 1^4 + 6^4 + 3^4 + 4^4$$
$$8208 = 8^4 + 2^4 + 0^4 + 8^4$$
$$9474 = 9^4 + 4^4 + 7^4 + 4^4$$

Please don't challenge your audience to find other such four-digit numbers, since to date no others have been found that have this property!

For a five-digit number, we get the following:

$$54,748 = 5^5 + 4^5 + 7^5 + 4^5 + 8^5$$

And then we can always take this pattern to an extremely large number such as

$$4,679,307, 774 = 4^{10} + 6^{10} + 7^{10} + 9^{10} + 3^{10} + 0^{10} + 7^{10} + 7^{10} + 7^{10} + 4^{10}$$

Using factorials, we can also set up similar arrangements, and it is believed that there are only four such examples as shown below:

$$1 = 1!$$
$$2 = 2!$$
$$145 = 1! + 4! + 5!$$
$$40,585 = 4! + 0! + 5! + 8! + 5!$$

## Splitting Numbers

We can even split numbers other than individual digits and still end up with some spectacular results such as the following:

$$1233 = 12^2 + 33^2$$
$$8833 = 88^2 + 33^2$$
$$5,882,353 = 588^2 + 2353^2$$
$$94,122,353 = 9412^2 + 2353^2$$
$$1,765,038,125 = 17650^2 + 38125^2$$
$$2,584,043,776 = 25840^2 + 43776^2$$

There are many more such examples, but we can also focus on a "reverse" situation, where we would take the difference of the squares of a split number rather than the sum of squares, such as with the following: We reverse the two parts so that from 48, we would take an reverse the split digits 8 and 4 and subtract the squares: $48 = 8^2 - 4^2$. Here are several more such examples.

3468 to be split as 34 and 68, so that $68^2 - 34^2 = 3{,}468$

16,128 to be split as 16 and 128, so that $128^2 - 16^2 = 16{,}128$

34,188 to be split as 34 and 188, so that $188^2 - 34^2 = 34{,}188$

216,513 to be split as 216 and 513, so that $513^2 - 216^2 = 216{,}513$

416,768 to be split as 416 and 768, so that $768^2 - 416^2 = 416{,}768$

2,661,653 to be split as 266 and 1653, so that
$$1653^2 - 266^2 = 2{,}661{,}653$$

59,809,776 to be split as 5980 and 9776,
so that $9776^2 - 5980^2 = 59{,}809{,}776$

There are many more such examples that your audience might feel compelled to search for. We wish them luck!

We can then take this even a step further in our effort to amaze the audience with the most unusual relationships. Consider displaying portions of numbers as the sum of cubes, as we show with a few examples here:

41,833, which we then split up as follows to get the sum of cubes: $4^3 + 18^3 + 33^3 = 41{,}833$

$$221{,}859 = 22^3 + 18^3 + 59^3$$
$$444{,}664 = 44^3 + 46^3 + 64^3$$
$$487{,}215 = 48^3 + 72^3 = 15^3$$
$$336{,}701 = 33^3 + 67^3 + 01^3$$
$$982{,}827 = 98^3 = 28^3 = 27^3$$
$$983{,}221 = 98^3 = 32^3 + 21^3$$
$$166{,}500{,}333 = 166^3 + 500^3 + 333^3$$

Once again this is not an exhaustive list and there are more such numbers that lend themselves to this unusual splitting arrangement.

We can always look for nice relationships between numbers. With some creativity we can establish another form of "friendliness" between numbers. Some of them can be truly mind-boggling! Take for example the pair of numbers: 6205 and 3869.

At first look, there seems to be no apparent relationship. But with some luck and imagination, we can get some fantastic results:

$$6205 = 38^2 + 69^2 \text{ and } 3869 = 62^2 + 05^2$$

We can even find another pair of numbers with a similar relationship. Consider these numbers.

$$5965 = 77^2 + 06^2 \text{ and } 7706 = 59^2 + 65^2$$

Imagine how your audience will feel when you expose this amazing relationship.

## An Unusual Number Property

There are times when number relationships produce the most incredible results. We offer here one such situation. If you multiply any number by a number consisting of all the same digits, then it is possible to create another number with all the same digits. The method is as follows: if you multiply a two-digit number by a number with all the same digits, say, five same digits, then separate this product by taking five digits (since we used a number with five same digits) from the right side of the newly found product and added to the remaining number. The result will be a number with all the same digits. For example, suppose we multiply the number 86 by the multi-digit number 44,444, so that we get 3,822,184. We then take five digits from the right side of this number, in this case 22,184 and added it to the remaining number as 38 + 22,184 = 22,222.

To further convince you, let's consider another situation. Supposing this time we multiply $1018 \times 888,888 = 904,887,984$. Since we began with a six-digit number, we will chop off the first six digits of our product, namely 88,7984 and added to the remaining number 904, we get: 887,984 + 904 = 888,888.

We can take this a step further by taking the square of the same-digits number, perhaps one of four digits, such as $2222^2 = 4,937,284$. We then take the last four digits and add it to the remaining number to get: 7284 + 493 = 7777.

A special situation is one we use the number $7777^2 = 60481729$. Now splitting the number as before, we get $6048 + 1729 = 7777$, which is the number we started with. There are some exceptions as $5555^2$. Here we have $5555 \times 5555 = 30,858,025$. When we split this number up by taking four digits from the right side of our product, and do the same addition as above, we obtain: $8025 + 3085 = 11,110$. Using a calculator and trying different combinations of these special multiplications could be quite enchanting, especially when the experimenters notice a pattern evolving.

## Friendly Numbers

There is a relationship, which has been named numbers as "friendly." What could possibly make two numbers friendly? Mathematicians have decided that two numbers are to be considered friendly (or as sometimes used in the more sophisticated literature, "amicable") if the sum of the proper divisors[2] (or factors) of one number equals the second number *and* the sum of the proper divisors of the second number equals the first number as well. Sounds complicated? It really isn't. Just take a look at the smallest pair of friendly numbers: 220 and 284.

The divisors (or factors) of **220** are 1, 2, 4, 5, 10, 11, 20, 22, 44, 55, and 110. Their sum is $1+2+4+5+10+11+20+22+44+55+110 = \mathbf{284}$.

The divisors of **284** are 1, 2, 4, 71, and 142, and their sum is $1 + 2 + 4 + 71 + 142 = \mathbf{220}$.

This shows that the two numbers can be considered *friendly numbers*.

The second pair of friendly numbers, which were discovered by the famous French mathematician Pierre Fermat (1601–1665), is 17,296 and 18,416.

In order for us to establish their friendliness relationship, we need to find all of their prime factors, which are: $17,296 = 2^4 \cdot 23 \cdot 47$, and $18,416 = 2^4 \cdot 1151$.

Then we need to create all the numbers from these prime factors as follows: The sum of the factors of 17,296 is

$$1 + 2 + 4 + 8 + 16 + 23 + 46 + 47 + 92 + 94 + 184 + 188 +$$
$$368 + 376 + 752 + 1081 + 2162 + 4324 + 8648 = \underline{18,416}$$

The sum of the factors of 18,416 is

$$1 + 2 + 4 + 8 + 16 + 1151 + 2302 + 4604 + 9208 = \underline{17,296}$$

Once again, we notice that the sum of the factors of 17,296 is equal to 18,416, and conversely, the sum of the factors of 18,416, is equal to 17,296. This qualifies them to be considered a pair of friendly numbers.

There are many more such pairs; so, for starters, here are a few more pairs of friendly numbers:

1184 and 1210

2,620 and 2,924

5020 and 5564

6,232 and 6,368

10,744 and 10,856

9,363,584 and 9,437,056

111,448,537,712 and 118,853,793,424

An ambitious audience might want to verify the above pairs' "friendliness!"

For the experts, the following is one method for finding pairs of friendly numbers:

Let $a = 3 \cdot 2^n - 1$, $b = 3 \cdot 2^{n-1} - 1$, and $c = 3^2 \cdot 2^{2n-1} - 1$, where $n$ is an integer greater than or equal to 2, and $a$, $b$, and $c$ are all prime numbers. It then follows that $2^n ab$ and $2^n c$ are friendly numbers. We should note, for $n = 2, 4,$ and 7, that $a$, $b$, and $c$ are all prime for $n$ less than or equal to 200.

Another form of friendliness can be seen with the following examples:

$$3869 = 62^2 + 05^2 = 6205 = 38^2 + 68^2$$
$$5965 = 77^2 + 06^2 = 7706 = 59^2 + 65^2$$

Are there other numbers that exhibit such friendliness?

We can even set up an analogous cycle using cubes:

Starting with 55: $5^3 + 5^3 = 250$, then 250: $2^3 + 5^3 + 0^3 = 133$, then 133: $1^3 + 3^3 + 3^3 = 55$, which is the number we started from. This can be done with other sequences of numbers such as:

$$136, 244, 136$$

$$919, 1,459, 919$$

$$160, 217, 352, 160$$

## The Magic of Square Numbers

Let's consider a certain "magic" of square numbers. First of all, let's take a slight detour to marvel at another curiosity. Sometimes peculiarities so simple can be interesting. Take for example the fact that there are only two numbers, 2 and 11, where their squares increased by 4 will yield a cube.

$$2^2 = 4, \text{ then by adding 4, we get } 4 + 4 = 8 = 2^3$$

$$11^2 = 121, \text{ then by adding 4, we get } 121 + 4 = 125 = 5^3$$

Now let's take a look at a list of square natural numbers and see if there is any pattern to be recognized. Patterns always seem to provide enrichment or enlightenment among an audience.

| $1^2$ | $2^2$ | $3^2$ | $4^2$ | $5^2$ | $6^2$ | $7^2$ | $8^2$ | $9^2$ | $10^2$ | $11^2$ | $12^2$ |
|---|---|---|---|---|---|---|---|---|---|---|---|
| **1** | **4** | **9** | 16 | 2**5** | 3**6** | 4**9** | 64 | 81 | 10**0** | 121 | 14**4** |

| $13^2$ | $14^2$ | $15^2$ | $16^2$ | $17^2$ | $18^2$ | $19^2$ | $20^2$ | $21^2$ |
|---|---|---|---|---|---|---|---|---|
| 16**9** | 19**6** | 22**5** | 256 | 28**9** | 32**4** | 36**1** | 40**0** | 44**1** |

One thing that may be quickly noticed among the square numbers listed is that the units digits, which we have bold underlined above, follow a specific pattern, namely, 1, 4, 9, 6, 5, 6, 9, 4, 1, 0, **1, 4, 9, 6, 5, 6, 9, 4, 1, 0**, 1,.... This pattern will continue without end. A clever person seeing this would be able to surmise that there are certain digits, which can never appear in the units-digit position, since they are missing from the repetitions list. That is, the digits 2, 3, 7 and 8 will never be the units digit of a square number. Furthermore, these numbers separated by the zero is a palindromic arrangement which can easily be spotted in the sequence: 1, 4, 9, 6, 5, 6, 9, 4, 1, 0, **1, 4, 9, 6, 5, 6, 9, 4, 1, 0**, 1,... and will continue ongoing.

There is probably no limit to the number of entertainments we can offer about square numbers. For example, the numbers 13 and 31 which we clearly see are reversals of one another, have, respectively, squares which are 169 in 961 which are also reversals of one another. Furthermore, if we take the product of these two numbers, we get $169 \times 961 = 162,409 = 403^2$, yet another square. If we want to take this a step further, the sum of the digits of 169 is $1 + 6 + 9 = 16 = 4^2$, and the sum of the digits of the square root of 169, which is 13, is $1 + 3 = 4$. To add to this amazingly beautiful relationship is another pair of numbers that have the same relationship:

These numbers are 12 and 21. If we follow the same pattern as we did with the numbers 13 and 31, we would get $12^2 = 144$ and $21^2 = 441$. The product of these two numbers is $144 \times 441 = 63,504 = 252^2$. In addition, $1+4+4 = 9=3^2$ and $1 + 2 = 3$; all analogous to the numbers 13 and 31.

While we are admiring square numbers, there are numbers, called *automorphic numbers*, whose squares end in the same digits, such as:

$$5^2 = 2\mathbf{5}$$
$$6^2 = 3\mathbf{6}$$
$$76^2 = 5,7\mathbf{76}$$
$$376^2 = 141,\mathbf{376}$$
$$625^2 = 390,\mathbf{625}$$
$$90,625^2 = 8,212,8\mathbf{90,625}$$
$$890,625^2 = 793,212,\mathbf{890,625}$$
$$1,787,109,376^2 = 3,193,759,921,\mathbf{787,109,376}$$
$$8,212,890,925^2 = 67,451,572,418,\mathbf{212,890,625}$$

After observing this pattern, the question is how can we create other such automorphic numbers?

Suppose we take the next-to-last automorphic number above and chop off several of their left side digits, so that we consider the number **921,787,109,376,** which when we square it, we get the number 849,691,475,011,761,**787,109,376**. You will notice that the last 10 digits are the same. This can be done with any of the above automorphic numbers, as we can also see with the previously calculated numbers: $90{,}625^2 = 8{,}212{,}\mathbf{890{,}625}$, and $890{,}625^2 = 793{,}212{,}\mathbf{890{,}625}$.

At this point, the audience may like to experiment taking on to the front of some of these numbers a few random digits keeping the terminal digits as shown above and finding that in each case automorphic numbers will be created. Keep in mind that at most there are two suffix groups of a specific number of digits that can be used to create automorphic numbers. For example, the numbers 625 and 376 are the only number of three digits that can be used to make automorphic numbers, such as the number $1{,}234{,}\mathbf{625}^2 = 1{,}524{,}298{,}890{,}\mathbf{625}$.

We should note that the number 90,625 is the only five-digit automorphic number. We can see a few uses of that five-digit automorphic number above.

Here are the automorphic numbers up to $10^{15}$:

1, 5, 6, 25, 76, 376, 625, 9376, 90625, 109376, 890625, 2890625, 7109376, 12890625, 87109376, 212890625, 787109376, 1787109376, 8212890625, 18212890625, 81787109376, 918212890625, 9918212890625, 40081787109376, 59918212890625, 259918212890625, 740081787109376

At this point your audience has lots to experiment with and to try to create numbers whose square ends up with the same end digits as the original number. Lots of fun lurks in the future!

## More Square-Number Patterns

Another entertaining relationship that can evolve from squared numbers is the following, which needs no further explanation.

$$10^2 + 11^2 + 12^2 = 13^2 + 14^2$$
$$21^2 + 22^2 + 23^2 + 24^2 = 25^2 + 26^2 + 27^2$$
$$36^2 + 37^2 + 38^2 + 39^2 + 40^2 = 41^2 + 42^2 + 43^2 + 44^2$$
$$55^2 + 56^2 + 57^2 + 58^2 + 59^2 + 60^2 = 61^2 + 62^2 + 63^2 + 64^2 + 65^2$$

This pattern continues and your audience should be thoroughly amazed at this pattern — notice that the number of terms on the right side is one less than the number of terms on the left side. Now the question might come up as to how does one find the first number of each line? We can use the formula $n(2n + 1)$, where $n$ is the number of terms on the right side of the equation. Therefore, to find the next such equation, we have $n = 6$, so that the first number will be $6(12 + 1) = 6 \times 13 = 78$, as shown here:

$$78^2 + 79^2 + 80^2 + 81^2 + 82^2 + 83^2 + 84^2$$
$$= 85^2 + 86^2 + 87^2 + 88^2 + 89^2 + 90^2$$

Ambitious audiences may wish to extend this further. In any case, the pattern could be quite entertaining as it is a bit mind-boggling!

## A Simple Pattern

On a simpler scale, this can also be done with first power numbers, such as shown here:

$$1 + 2 = 3$$
$$4 + 5 + 6 = 7 + 8$$
$$9 + 10 + 11 + 12 = 13 + 14 + 15$$
$$16 + 17 + 18 + 19 + 20 = 21 + 22 + 23 + 24$$

The pattern continues and those with some insight will notice why this will hold true continuously.

By the way, one of the numbers shown here 3334 taken to the third power also provides some entertainment, in that $3334^3 = 37,059,263,704$, and if we break that number up into three parts and take the sum, surprisingly arrive at: $370 + 5926 + 3704 = 10,000$. Cute!

## An Unexpected Pattern

Patterns in mathematics tend to crop up when we would least expect it. We offer the following example of how a rather innocuous series of numbers when squared result in a totally unexpected pattern.

$$4^2 = 16$$
$$34^2 = 1156$$
$$334^2 = 111,556$$
$$3334^2 = 11,115,556$$
$$33334^2 = 1,111,155,556$$
$$333334^2 = 111,111,555,556$$

Although the clever reader will probably be able to continue this pattern, it would be pleasantly challenging that's great. Fixing your hernia is also fixing your hernia has also made you more responsive we must get Russell to sign this before the end of next week To seek other patterns of a similar kind.

## A Peculiarity of the Number 48

If we consider the proper factors of a number to include all the factors excluding 1 and the number itself, then we can say that the fourth power of the number 48 is a number, which is equal to the product of all of its proper factors. This can probably be best demonstrated on a calculator as follows: the proper factors of the number 48 are 2, 3, 4, 6, 8, 12, 16, and 24. Thus, we have: $2 \times 3 \times 4 \times 6 \times 8 \times 12 \times 16 \times 24 = 5,308,416 = 48^4$.

Another cute number property is one where the sum of all the divisors is equal to a perfect square. You can either show your audience some of these and ask them to find others, or simply have them prove if that is true for the following numbers: 3, 22, 66, 70, 81, and others. Let's use, for an example, the number 66. The sum of its divisors is: $1 + 2 + 3 + 6 + 11 + 22 + 33 + 66 = 144 = 12^2$.

Another example can be shown for the number 22, where the sum of its factors $1 + 2 + 11 + 22 = 36 = 6^2$.

## Prime Numbers:

Prime numbers are numbers that have only two different divisors, that is 1 and the number itself. Because the divisors have to be different, the number 1 is not considered a prime number, since it only has one divisor the number 1 itself. Prime numbers provide fascinating some fascinating aspects of our number system and can be used nicely to entertain folks in a simple way. Here are the prime numbers less than 1000:

| | | | | | | | | | | | | | | | | |
|---|---|---|---|---|---|---|---|---|---|---|---|---|---|---|---|---|
| 2 | 3 | 5 | 7 | 11 | 13 | 17 | 19 | 23 | 29 | 31 | 37 | 41 | 43 | 47 | 53 | 59 61 |
| 67 | 71 | 73 | 79 | 83 | 89 | 97 | 101 | 103 | 107 | 109 | 113 | 127 | 131 | 137 | 139 | |

149 151 157 163 167 173 179 181 191 193 197 199 211 223  227
229 233 239 241 251 257 263 269 271 277 281 283 293 307  311
313 317 331 337 347 349 353 359 367 373 379 383 389 397  401
409 419 421 431 433 439 443 449 457 461 463 467 479 487  491
499 503 509 521 523 541 547 557 563 569 571 577 587 593  599
601 607 613 617 619 631 641 643 647 653 659 661 673 677  683
691 701 709 719 727 733 739 743 751 757 761 769 773 787  797
809 811 821 823 827 829 839 853 857 859 863 877 881 883  887
907 911 919 929 937 941 947 953 967 971 977 983 991 997.s

There are also rather unusual-looking prime numbers such as the number: 90909090909090909090909090909091, which is comprised of fourteen 90's ending with 91.

You can also have fun with prime numbers (those that have only have themselves and 1 as factors). For example, 113 is a prime number, since its only factors are 113 and 1. However, this particular prime is the smallest prime number where all the arrangements of the digits are also a prime numbers — that is, 131 and 311. Other such primes are 337 and 199.

## How to Establish a Prime Number

There are times when your audience will appreciate a clever technique for doing something that might well have been taught to them in high school, but clearly was not. That is, how to establish whether a number is a prime number or not. Before we set up a technique for establishing prime numbers, we need to review that $n! = 1 \times 2 \times 3 \times 4 \times 5 \times 6 \times \cdots \times n$. The rule for establishing a prime is that if $n! + 1$ is divisible by $n + 1$, then $n + 1$ is a prime number. Suppose we would like to test this to see if the number 11 is a prime number. Therefore, we say that $11 = n + 1$, whereupon $n = 10$. We now seek 10!, which is equal to 3,628,800. Thus, $3,628,800 + 1 = 3,628,801 = 11 \times 329,891$. Therefore, we can conclude that 11 is a prime number.

## A Strange Coincidence

Can you imagine that the first 6 prime numbers could be divisors of 6 consecutive numbers. Well, there is such a case, and it could be a curiosity which can be quite entertaining. The consecutive numbers, 788, 789, 790, 791, 792, and 793, are divisible by 2, 3, 5, 7, 11, and 13, respectively. There you have it!

## Prime Numbers in 10-Periods

To begin this unit, we will define a 10-period as a sequence of 10 consecutive numbers such as 1–10, or 11–20, or 21–30, or 31–40, etc. In most of these 10-periods the number of prime numbers to be found is usually 2 or 3 primes, such as in the 10-period 41–50, which has 3 prime numbers are 41, 43, and 47.

While in the 10-period 21–30 there are only 2 prime numbers, namely, 23 and 29. It is rather rare that there are 4 prime numbers in any one of the 10-periods going forward. Naturally, in the first

10-period, there are 4 primes, which are 2, 3, 5, and 7. Also in the next 10-period, there are also 4 prime numbers, which are 11, 13, 17, and 19. We have to go long way to get the next 10-period which has 4 prime numbers and that is the 10-period 101–110, where the prime numbers are 101, 103, 107, and 109. Then we have to go quite a bit further to get the next 10-period span to find 4 prime numbers and that is the 10-period 821–830, which contains the prime numbers 821, 823, 827, and 829. Your audience may be interested to know when the next 10-periods occur that contain 4 prime numbers. In Figure 1.7, we show the next 5 such 10-periods.

| 10-period | Prime numbers in the 10-period |
|-----------|--------------------------------|
| 1481–1490 | 1481, 1483, 1487, 1489 |
| 1871–1880 | 1871, 1873, 1877, 1879 |
| 2081–2090 | 2081, 2083, 2087, 2089 |
| 3251–3260 | 3251, 3253, 3257, 3259 |
| 3461–3470 | 3461, 3463, 3467, 3469 |

**Figure 1.7**

It is noteworthy, that there are no other such 10-periods that contain 4 prime numbers. The question might come up what causes this restricted number of primes in one such period? Let's take a moment to inspect what the possible candidates there are in each of these intervals. First of all, there are 5 even numbers and only the number 2 of the even numbers can be considered a prime number. That leaves 5 other candidates for being a prime number, with the exception of the first interval. Of the 5 remaining odd numbers, the multiple of 5 must be eliminated except in the first interval. That leads the units digit of the remaining 4 numbers to be 1, 3, 7, and 9. However, there are times when some of these numbers are divisible by 3 or by 7, which would eliminate them from the prime-number category, such as the numbers 21, 93, 27, and 49. An interested audience might want to search for further 10-periods that contain 4 prime numbers.

# Twin Primes

Prime numbers can also be seen by their place on the list of primes. When 2 prime numbers have a difference of 2, they are considered *twin primes*. It is suspected that there are an infinite number of twin primes, but this has never been proved or disproved. The first few twin primes are: (3, 5), (5, 7), (11, 13), (17, 19), (29, 31), (41, 43), (59, 61), (71, 73), (101, 103), (107, 109), (137, 139), $\cdots$ , where we notice that 5 is the only number that will appear twice in the list of twin primes. The audience may be curious to know what the largest twin prime pair that has been discovered to date. As of September 2018, the largest twin prime pair is $(2996863034895 \times 2^{1290000} - 1, 2996863034895 \times 2^{1290000} + 1)$. There are 808,675,888,577,436 twin prime pairs less than $10^{18}$.

If you have properly motivated your audience, someone may ask is there a general format for expressing twin prime pairs. The answer is, with the exception of the first twin prime pair, namely, (3, 5), all others can be expressed in the form of $(6n - 1)$, $(6n + 1)$, where *n* is a natural number. A clever observer will also notice that every number between a pair of twin primes will be a multiple of 6, which can be easily substantiated with the first several twin primes.

# Some Prime Denominator Surprises

Remember that prime numbers have no factors other than themselves and 1. We now consider prime numbers (excluding 2 and 5), which are in the denominator of a fraction, and where the decimal expansion yields an *even* number of repetition digits. This will enable us to demonstrate some rather unbelievable relationships that will surely impress the audience. Let's consider a few fractions with an even number of repetition digits (the bar above the digits indicates that they repeat indefinitely):

$$\frac{1}{7} = 0.142857142857142857... = \overline{0.142857}$$

$$\frac{1}{11} = 0.090909... = \overline{0.09}$$

$$\frac{1}{13} = 0.076923076923... = \overline{0.076923}$$

$$\frac{1}{17} = \overline{0.0588235294117647}$$

$$\frac{1}{19} = \overline{0.052631578947368421}$$

$$\frac{1}{23} = \overline{0.0434782608695652173913}$$

We will now show this most unanticipated trick move. We will treat each one of these repeating periods as a number, and split this even-digit sequence of digits into two equal parts. When we add these two numbers, we get a very surprising number consisting on only 9s. This should truly amaze your audience with this unexpected relationship.[3] See Figure 1.8.

| $\frac{1}{7} = 0.142857142857142857... = \overline{0.142857}$ | 142<br>857<br>999 |
|---|---|
| $\frac{1}{11} = 0.090909... = \overline{0.09}$ | 0<br>9<br>9 |
| $\frac{1}{13} = \overline{0.076923}$ | 076<br>923<br>999 |
| $\frac{1}{17} = \overline{0.0588235294117647}$ | 05882352<br>94117647<br>99999999 |
| $\frac{1}{19} = \overline{0.052631578947368421}$ | 052631578<br>947368421<br>999999999 |
| $\frac{1}{23} = \overline{0.0434782608695652173913}$ | 04347826086<br>95652173913<br>99999999999 |

**Figure 1.8**

The awe that this result brings may entice the audience to seek further examples — just a further generation of a "wonder" that is hidden in mathematics. Needless to say, the results of these additions will truly impress your audience, and also foreshadow the next topic — Palindromes, which each of these sums are.

## Palindromic Numbers

There are certain categories of numbers that have particularly strange characteristics, which can truly entertain an audience. Here we consider numbers that read the same in both directions: left to right, or right to left. These are called *palindromic numbers*. First note that a palindrome can also be a word, phrase, or sentence that reads the same in both directions. Figure 1.9 shows a few amusing palindromes.

A
EVE
RADAR
REVIVER
ROTATOR
LEPERS REPEL
MADAM I'M ADAM
STEP NOT ON PETS
DO GEESE SEE GOD
PULL UP IF I PULL UP
NO LEMONS, NO MELON
DENNIS AND EDNA SINNED
ABLE WAS I ERE I SAW ELBA
A MAN, A PLAN, A CANAL, PANAMA
A SANTA LIVED AS A DEVIL AT NASA
SUMS ARE NOT SET AS A TEST ON ERASMUS
ON A CLOVER, IF ALIVE, ERUPTS A VAST, PURE EVIL; A FIRE VOLCANO

**Figure 1.9**

A palindrome in mathematics would be a number such as 666, or 123321 that reads the same in either direction. For example, the first five powers of 11 are palindromic numbers:

$$11^0 = 1$$

$$11^1 = 11$$

$$11^2 = 121$$
$$11^3 = 1331$$
$$11^4 = 14641$$

Once again, using a calculator, we find that there are some unusual occurrences which result from squaring numbers consisting of all 1s, which are numbers often called *reunits*, as we show below — another unexpected astonishment to observe. They result in palindromic numbers.

$$11^2 = 121$$
$$111^2 = 1331$$
$$1111^2 = 1234321$$
$$11111^2 = 123454321$$
$$111111^2 = 12345654321$$
$$1111111^2 = 1234567654321$$
$$11111111^2 = 123456787654321$$
$$111111111^2 = 12345678987654321$$

A small curiosity that should interest an audience is that the smallest number with an even number of digits beyond the reunit numbers whose square is a palindrome is the number 798,644, since $798,644^2 = 637,832,238,736$, which is a palindrome.

Now for the entertaining aspect of palindromic numbers. Here, we have a procedure to see how a palindromic number can be generated from a given number. All you need to do is to continually add a given number to its reversal (that is, the number written in the reverse order of digits) until you arrive at a palindrome. For example, a palindrome can be reached with a single addition such as with the starting number 23: the sum $23 + 32 = 55$, a palindrome.

Or it might take two steps, such as with the starting number 75: the two successive sums are

$75 + 57 = 132$ and $132 + 231 = 363$, which led us to a palindrome.

Or it might take three steps, such as with the starting number 86:

$$86+68=154, \quad 154+451=605, \quad 605+506=1111.$$

The starting number 97 will require 6 steps to reach a palindrome; while the starting number 98 will require 24 steps to reach a palindrome. It is important to be honest with your audience and caution them about using the starting number 196; as this one has not yet been shown to produce a palindrome number — even with over three million reversal additions. We still do not know if this one will ever reach a palindrome.

There are a few quirky results in this procedure. If you were to try to apply this procedure on 196, you would eventually — at the 16th addition — reach the number 227574622. Yet, amazingly, you would also reach that same sum at the 15th step of the attempt to get a palindrome from the starting number 788. This would then tell you that applying the procedure to the number 788, also has never been shown to reach a palindrome. As a matter of fact, among the first 100,000 natural numbers, there are 5996 numbers for which we have not yet been able to show that the procedure of reversal additions will lead to a palindrome. Some of these non-palindrome results are 196, 691, 788, 887, 1675, 5761, 6347, and 7436.

Now that you have motivated your audience, you might want to take this to another level by showing some unusual aspects in this process. For example, using this procedure of reverse and add, we find that some numbers yield the same palindrome in the same number of steps, such as 554, 752, and 653, which all produce the palindrome 11011 in 3 steps. In general, all integers in which the corresponding digit pairs symmetric to the middle 5 have the same sum, will produce the same palindrome in the same number of steps. The above three sample numbers, 554, 752, and 653, have this characteristic since the pair of digits on either side of the middle 5 have the same sum namely 9.

There are other integers that produce the same palindrome, yet with a different number of steps, such as the number 198, which, with repeated reversals and additions, will reach the palindrome 79497 in

5 steps, while the number 7299 will reach this same number in 2 steps, that is: $7299 + 9927 = 79497$.

To further amaze your audience, you can show them how we can determine the number of additions that we will have to do to reach a palindrome using this procedure. This takes the audience a bit further and will be, perhaps, more enlightening than entertaining. For a two-digit number $ab$ with digits $a \neq b$, the sum $a + b$ of its digits determines the number of steps needed to produce a palindrome. Clearly, if the sum of the digits less than 10, then only one step will be required to reach a palindrome, for example, $25 + 52 = 77$. If the sum of the digits is 10, then $ab + ba = 110$, and $110 + 011 = 121$, and 2 steps will be required to reach the palindrome. The number of steps required for each of the two-digit sums 11, 12, 13, 14, 15, 16, and 17, to reach a palindromic number are 1, 2, 2, 3, 4, 6, and 24, respectively.

Now we can take this to another level to appreciate some unusual aspects of palindromic numbers. We can arrive at some lovely patterns when dealing with palindromic numbers. For example, some palindromic numbers, when squared, also yield a palindrome. For example, $22^2 = 484$ and $212^2 = 44944$. On the other hand, there are also some palindromic numbers, when squared, do not yield a Palindromic number, such as $545^2 = 297{,}025$. Of course, there are also non-palindromic numbers that, when squared, yield a palindromic number, such as $26^2 = 676$, and $836^2 = 698{,}896$. These are just some of the entertainments that numbers provide. You may want to search for other such curiosities.

## Taking Palindromic Numbers Further

There are also some palindromic numbers that, when cubed, yield again palindromic numbers.

To this group belong all numbers of the form $n = 10^k + 1$, for $k = 1$, 2, 3, ... . When $n$ is cubed, it yields a palindromic number, which has $k - 1$ zeros between each consecutive pair of 1, 3, 3, 1, as we can see with the following examples:

$k = 1, n = 11:$    $11^3 = 1331$

$k = 2, n = 101:$    $101^3 = 1030301$

$k = 3, n = 1001:$    $1001^3 = 1003003001$

$k = 7, n = 10000001:$    $10000001^3 = 1000000300000030000001$

We can continue to generalize to further entertain the audience by getting some interesting patterns, such as when $n$ consists of three 1's and any even number of zeros symmetrically placed between the end 1's when cubed will give us a palindrome, such as

$$111^3 = 1367631$$
$$10101^3 = 1030607060301$$
$$1001001^3 = 1003006007006003001$$
$$100010001^3 = 1000300060007000600030001$$

Taking this even another step further, we find that when $n$ consists of four 1's and zeros in a palindromic arrangement, where the places between the 1's do not have same number of zeros, then $n^3$ will also be a palindrome, as we can see with the following examples:

$$11011^3 = 1334996994331$$
$$10100101^3 = 1030331909339091330301$$
$$10010001001^3 = 1,003,003,301,900,930,390,091,033,003,001$$

However, when the same number of zeros appears between the ones then the cube of the number will *not* result in a palindrome, as in the following example: $1010101^3 = 1030610121210060301$. As a matter of fact, the number 2201 is the only non-palindromic number, which is less than 280,000,000,000,000, and, when cubed, yields a palindrome: $2201^3 = 10662526601$.

Just for more amusement consider the following pattern with palindromic numbers:

$$12321 = \frac{333 \cdot 333}{1+2+3+2+1}$$

$$1234321 = \frac{4444 \cdot 4444}{1+2+3+4+3+2+1}$$

$$123454321 = \frac{55555 \cdot 55555}{1+2+3+4+5+4+3+2+1}$$

$$12345654321 = \frac{666666 \cdot 666666}{1+2+3+4+5+6+5+4+3+2+1}$$

and so on.

An ambitious reader may search for other patterns involving palindromic numbers.

## Prime Divisibilities

The bit earlier we discussed prime numbers, and we also presented special properties of the two important numbers in our number system, which are situated on either side of our base 10, namely the numbers 9 and 11. Part of our discussion of these two numbers showed how we can determine when a given number is divisible by either of these two special numbers. However, it can be very impressive and entertaining to an audience to show them how you can determine by inspecting a given number whether it is divisible by a prime number. Since we have already considered the earlier prime numbers where divisibility is concerned, the next in line would be the prime number 7. In order to determine if a given number is divisible by 7 we would invoke the following technique and then beyond that see how this can be used to discover divisibility rules for other prime numbers. The method that we will use to determine if a given number is divisible by 7 is as follows:

> *Delete the last digit from the given number, and then subtract twice this deleted digit from the remaining number. If the result is divisible*

*by 7, then the original number is divisible by 7. This process may be repeated until the result can be determined by simple inspection for divisibility by 7.*

Let's try one as an example to see how this rule works. Suppose we want to test the number 680,715 for divisibility by 7. Begin with 680,715 and delete its units digit, 5, and subtract its double, 10, from the remaining number: $68,071 - 10 = 68,061$. Since we cannot yet visually inspect the resulting number for divisibility by 7, we continue the process with the resulting number 68,061 and delete its units digit, 1, and subtract its double, 2, from the remaining number; we get: $6,806 - 2 = 6,804$. This did not bring us any closer to visually being able to check for divisibility by 7; therefore, we continue the process with the resulting number 6804 and delete its units digit, 4, and subtract its double, 8, from the remaining number; we get: $680 - 8 = 672$. Since we still cannot visually inspect the resulting number, 672, for divisibility by 7, we continue the process with the resulting number 672 and delete its units digit, 2, and subtract its double, 4, from the remaining number we get: $67 - 4 = 63$, which we can easily see is divisible by 7. Therefore, the original number 680,715 is also divisible by 7.

Before continuing with our discussion of divisibility of prime numbers it would be wise to practice this technique with a few randomly selected numbers and then check the results with a calculator.

To justify this procedure of determining divisibility by 7, consider the various possible terminal digits (that you are "dropping") and the corresponding subtraction that is actually being done by dropping the last digit. In the chart (Figure 1.10), you will see how dropping the terminal digit and doubling it, the number being subtracted gives us in each case a multiple of 7. That is, we have taken "bundles of 7" away from the original number. Therefore, if the remaining number is divisible by 7, then so is the original number, because you have separated the original number into two parts, each of which is divisible by 7, and therefore, the entire number must be divisible by 7.

| Terminal digit | Number subtracted from original | Terminal digit | Number subtracted from original |
|---|---|---|---|
| 1 | $20 + 1 = 21 = \ 3 \times 7$ | 5 | $100 + 5 = 105 = 15 \times 7$ |
| 2 | $40 + 2 = 42 = \ 6 \times 7$ | 6 | $120 + 6 = 126 = 18 \times 7$ |
| 3 | $60 + 3 = 63 = \ 9 \times 7$ | 7 | $140 + 7 = 147 = 21 \times 7$ |
| 4 | $80 + 4 = 84 = 12 \times 7$ | 8 | $160 + 8 = 168 = 24 \times 7$ |
| | | 9 | $180 + 9 = 189 = 27 \times 7$ |

**Figure 1.10**

We should be able to use the technique that we just developed for determining divisibility by 7 to create divisibility-determining techniques for other prime numbers. The next prime number to be considered would be the number 13. The technique for divisibility by 13 is as follows:

> *The technique for determining if a number is divisible a 13 is the similar to the rule for testing divisibility by 7, except that the 7 is replaced by 13, and instead of subtracting twice the deleted digit, we subtract nine times the deleted digit each time.*

Perhaps it would be best to consider an example where we seek to check for divisibility by 13 for the number 9776. We begin by deleting its units digit, 6, and then subtract $9 \times 6 = 54$, from the remaining number: $977 - 54 = 923$. Since we still cannot visually inspect the resulting number 923 for divisibility by 13, we continue the process with the resulting number 923 and delete its units digit, 3, and subtract 9 times this digit ($9 \times 3 = 27$) from the remaining number: $92 - 27 = 65$, which is divisible by 13, and therefore, the original number, 9776, is also divisible by 13.

Now we might want to see how we determined the "multiplier" 9 in our trick. We sought the smallest multiple of 13 that ends in a 1. That was 91, where the tens digit is 9 times the units digit. Once again consider the various possible terminal digits and the corresponding subtractions in Figure 1.11.

| Terminal digit | Number subtracted from original | Terminal digit | Number subtracted from original |
|---|---|---|---|
| 1 | $90 + 1 = 91 = 7 \times 13$ | 5 | $450 + 5 = 455 = 35 \times 13$ |
| 2 | $180 + 2 = 182 = 14 \times 13$ | 6 | $540 + 6 = 546 = 42 \times 13$ |
| 3 | $270 + 3 = 273 = 21 \times 13$ | 7 | $630 + 7 = 637 = 49 \times 13$ |
| 4 | $360 + 4 = 364 = 28 \times 13$ | 8 | $720 + 8 = 728 = 56 \times 13$ |
|  |  | 9 | $810 + 9 = 819 = 63 \times 13$ |

**Figure 1.11**

In each case, a multiple of 13 is being subtracted one or more times from the original number. Hence, if the remaining number is divisible by 13, then the original number is divisible by 13.

If the previous techniques are presented properly, your audience should be properly motivated and curious to see if they can determine a technique for testing divisibility of a given number by 17. We offer that here as follows:

*Delete the units digit and subtract 5 times the deleted digit each time from the remaining number until you reach a number small enough to visually determine its divisibility by 17.*

We justify the technique for divisibility by 17 as we did those for 7 and 13. Each step of the process requires us to subtract a "bunch of 17s" from the original number until we reduce the number to a manageable size by which we can make a visual inspection of divisibility by 17. This time we can see that the multiplier is 5, since we will be deducting multiples of 17, such as 51, 102, 153, and so on, from the original number.

The patterns developed in the preceding three divisibility techniques (for 7, 13, and 17) should enable you to produce analogous ones for testing divisibility by larger primes. Figure 1.12 presents the "multipliers" of the deleted digits for various primes.

| To test divisibility by | 7 | 11 | 13 | 17 | 19 | 23 | 29 | 31 | 37 | 41 | 43 | 47 |
|---|---|---|---|---|---|---|---|---|---|---|---|---|
| Multiplier | 2 | 1 | 9 | 5 | 17 | 16 | 26 | 3 | 11 | 4 | 30 | 14 |

**Figure 1.12**

You may want to have this chart extended. It's fun, and it will increase your audience's perception of mathematics. You may also want to have your audience consider divisibility rules to include composite (i.e., non-prime) numbers. Why the following rule refers to relatively prime factors and not just any factors is something that will sharpen their understanding of number properties. Perhaps the easiest response to this question is that relatively prime factors have independent divisibility rules, whereas other factors may not. Therefore, your audience may wish to consider the following technique for divisibility by composite numbers:

*A given number is divisible by a composite number, if it is divisible by each of its relatively prime factors.*

(Two numbers are relatively prime if they have no common factors other than 1.) Figure 1.13 offers illustrations of this rule.

| To be divisible by | 6 | 10 | 12 | 15 | 18 | 21 | 24 | 26 | 28 |
|---|---|---|---|---|---|---|---|---|---|
| The number must be divisible by these relatively-prime factors | 2, 3 | 2, 5 | 3, 4 | 3, 5 | 2, 9 | 3, 7 | 3, 8 | 2, 13 | 4, 7 |

**Figure 1.13**

At this juncture your audience has not only a rather comprehensive list of techniques for testing divisibility, but also an interesting insight into elementary number theory. They may be encouraged to practice using these rules (to instill greater familiarity) and try to develop techniques to test divisibility by other numbers in base 10 and to generalize these to numbers in other bases. Unfortunately, lack of space prevents a more detailed development here. Yet, we hope to have enabled you to whet the audience's appetite regarding divisibility. Remember, although this is a rather extensive development, it must be presented in a fashion that will enchant the audience and that depends on the nature of the participants.

## Guessing the Missing Number

Guessing a number always catches the audiences' curiosity. Begin by presenting the following three numbers: 1, 3, and 8 and asks the audience for a fourth number *n,* such that when the product of *any pair* of the four numbers is added to 1, the result will be a square number. In other words, if we take the product of 2 of the given numbers, say, $3 \times 8 = 24$ and when added to 1, we get 25, which is a square number. Or, another possibility would be $8 \times 1 = 8$ and when added to 1, we get 9, which is also square number. Typically, the audience will start trying to replace the number *n* with smaller numbers and probably find that their selected numbers don't hold this pattern for every pair from four numbers being considered.

Here we provide the answer: $n = 120$, since $1 \times 120$, when added to $1 = 121$, which is $11^2$, and $3 \times 120 = 360$ and when added to 1 is $361 = 19^2$, and $8 \times 120$ when added to $1 = 961 = 19^2$. Although this may take some time on the part of the audience, this could be entertaining, since the sought-after number is quite far from the other given numbers.

## An Arithmetic Phenomenon

At first, this activity will enchant the audience, and then (if properly presented) it will make them wonder why this result works as it does. This is a wonderful opportunity to show off the usefulness of algebra, for it will be through algebra that their curiosity will be quenched. Present the following to the audience.

Select any three-digit number with all digits different from one another. Write all possible two-digit numbers that can be formed from the three digits selected. Then divide their sum by the sum of the digits in the original three-digit number. Everyone in the audience should all get the same answer, 22.

This ought to elicit a big "WOW!" Let's consider the three-digit number 365. Take the sum of all the possible two-digit numbers that can be formed from these three digits. $36 + 35 + 63 + 53 + 65 + 56 = 308$. Then we get sum of the digits of the original number, which is $3 + 6 + 5 = 14$. When we are to divide 308 by 14 we get 22, which everyone should have gotten regardless of which original three-digit number was selected.

Let's analyze this unusual result, where everyone arrived at the number 22, regardless of which three-digit number they started with. We will begin with a general representation of the selected number: $100x + 10y + z$. We now take the sum of all the two-digit numbers taken from the original three digits:

$$(10x + y) + (10y + x) + (10x + z) + (10z + x) + (10y + z) + (10z + y)$$
$$= 10(2x + 2y + 2z) + (2x + 2y + 2z)$$
$$= 11(2x + 2y + 2z)$$
$$= 22(x + y + z)$$

When this value $22(x + y + z)$ is divided by the sum of the digits, $(x + y + z)$, the result is 22.

With this algebraic explanation, your audience ought to get a genuine appreciation as to how nicely algebra allows us to understanding arithmetic curiosities. Once again, we see how algebra allows us to explain simple arithmetic phenomenon, and also exhibit its beauty.

## Baffling Your Audience with a Seemingly Tough Challenge

When asked to divide 59 by 10 it is clear that we get 5 with a remainder of 9. Now ask your audience to find a number, which when divided by 10 leaves a remainder of 9, and when divided by 9 will leave a remainder of 8. And when it is divided by 8 it will leave a remainder of 7. Then moving along till when this number is divided by 3, it will leave a remainder of 2, and finally, when divided by 2 it will leave a remainder of 1. The challenge for your audience is to find a number that has these characteristics. At first glance this seems to be a rather insurmountable task. Of course, you can impress your audience by telling them that one such number is 14,622,042,959, and they can verify that using a calculator. However, it is unrealistic to expect the audience to come up with this number. Therefore, you can tell them that there is also a small number that would have the sought-after characteristics mentioned above. One such smaller number is 3,628,799, which also satisfies the original challenge. Yet, the smallest number that meets the above criteria can be obtained by looking for the least common multiple of the numbers 1, 2, 3, 4, ... ,8, 9, 10, which is $2^3 \times 3^2 \times 5 \times 7 = 2520$ and then subtract 1 to get 2519, which is smallest number that satisfies the criteria requested.

## Moving the Units Digit to the Front Position

Here again, a calculator will be very useful so as to allow focus on the product and not on the process of multiplication. You can entertain folks very easily by showing them that certain numbers when multiplied by 4 moves the units digit to the first position as shown with the following examples: $102,564 \times 4 = 410,256$.

We can take this a step further by repeating the six-digit number (102,564) to make a 12-digit (102,564,102,564) number and the same thing will happen: $102,564,102,564 \times 4 = 410,256,410,256$.

We could even extend this further by using this six-digit number and repeating it 3 times to get the following 18-digit number 102,564,102,564,102,564 and the multiplying it by 4 to get 410,256,410,256,410,256. Each time the units digit moves to the front of the number.

This can be done with other multiples of 4, for example: $179,487 \times 4 = 717,948$, and for the six-digit number: $179,487,179,487 \times 4 = 717,948,717,948$.

The following numbers also lend themselves to this curious property.

$$128,205 \times 4 = 512,820$$
$$153,846 \times 4 = 615,384$$
$$205,128 \times 4 = 820,512$$
$$230,769 \times 4 = 923,076$$

The question then arises can this be done with multiples other than 4. With the multiple 5 you can see in the following example the same procedure as that above:

$$142,85\underline{7} \times 5 = \underline{7}14,285, \text{ and for the six-digit version:}$$
$$142,857,142,85\underline{7} \times 5 = \underline{7}14,285,714,285.$$

An ambitious reader may search for other numbers that exhibit this unusual property. There are a number of ways of getting this multiplier phenomena. One such uses the formula $\frac{n}{10n-1}$. Suppose we let $n = 2$, then the number we would use is $\frac{2}{20-1} = \frac{2}{19}$. Beginning with $\frac{1}{19} = 0.05263157894736842\underline{1}$ (which then repeats), and multiplying this number by 2, we get $0.\underline{1}05263157894736842$; once again, notice how the units digit was moved to the front position. When we multiply $0.\underline{05263}1578947368421$ by 3, we get $0.157894736842\underline{105263}$; now notice the shift of five digits from the front position to the rear position. However, if we follow the original formula, namely, $\frac{n}{10n-1}$, we will stay with the original movement of the units digit to the front position. Just to help solidify this procedure, we will apply it once more where $n = 7$. Thus, we now have $\frac{7}{70-1} = \frac{7}{69}$ and $\frac{1}{69} = 014492753623188405797\underline{1}$ (which again repeats). Then we seek $7 \times 014492753623188405797\underline{1} = \underline{1},014,492,753,623,188,405,797$, and we notice that the units digit wandered over to the front position. If we now multiply this last-obtained number $1,014,492,753,623,188,405,79\underline{7}$ by 7, we get $\underline{7},101,449,275,362,318,840,579$, where once again the units digit wandered over to the front position. For further entertainment, this can be continued each time multiplying by 7 and finding that the units digit moves to the front position.

Naturally, instead of multiplying by 7 we can take one of these numbers and divide them by 7, noticing that the movement of the front position number would be relegated to the units digit.

## Unique Digit Peculiarities

The representation of all nine digits often fascinates the observer. Let's consider some novelties that occur when we use our 10 digits.

For example, if we square the number 99066 — which in itself is a curious number, since you can flip it over and it maintains its same value — we get the largest square number that can be formed using our 10 digits exactly once. The number is 9,814,072,356. Symbolically we have $\sqrt{9,814,072,356} = 99066$, or stated another way, $99066^2 = 9,814,072,356$.

Another curiosity that can be entertaining is to ask your audience to see if they can form two five-digit numbers using all of our 10 digits exactly once and which yield the largest possible product. Here we would want 2 numbers that are close to each other in value and where we assigned the digits alternatingly in descending order. This yields the following two numbers 96,420 and 87,531, whose product is then 8,439,739,020.

You may want to be "cute" to challenge your audience to determine what makes the following number unusual and where the answer will typically find them completely off-guard. Ask them what property the following number has: 8,549,176,320. Please don't let them struggle too long to figure this out, because in comparison to everything else we do in this book, this one is rather silly. The simple truth about this number is that the digits are arranged in alphabetical order according to their English names. Sorry for the silliness — which sometimes can also get a chuckle.

One rather unexpected result happens when we subtract the symmetric numbers consisting of the digits in consecutive reverse-order and in numerical order: 987,654,321 − 123,456,789 to get 864,197,532. This symmetric subtraction used each of the 9 digits exactly once in each of the numbers being subtracted, and surprisingly, resulted in a difference that also used each of the nine digits exactly once. We can also do this with the 0 included: 9,876,543,210 − 0123,456,789 = 9,753,086,421. If we leave out the 3 and 6, we can also get a difference which has all different digits — excluding the 3 and 6, as with this: 98,754,210 − 01,245,789 = 97,508,421. Notice that no digit is repeated in all 3 numbers of the subtraction example, and the 2 numbers being subtracted are in ascending or descending order. We can also take a three-digit number where the digits are in ascending and descending order and the same thing with be true, as with: 954 − 459 = 495. By the way, all of the above subtractions are unique.

While on the topic of using all 10 digits in arithmetic, consider the two numbers 96,702 and 58,413. Together these two numbers use all the digits in our number system exactly once. If we now take the square of each of these numbers $96,702^2 = 9,351,276,804$, and $58,413^2 = 3,412,078,569$, we see that each of these squares use all of our 10 digits exactly once. There are 3 other pairs of numbers that share this unique characteristic and they are 35,172 and 60,984 together use all 10 digits and each of their squares also use all 10 digits: $35,172^2 = 1,237,069,584$, and $50,984^2 = 3,719,048,256$.

Similarly, 59,403 and 76,182 together use all the digits exactly once, and their squares do the same: $59,403^2 = 3,528,716,409$ and $76,182^2 = 5,803,697,124$. Remember, each of the squares represented is comprised of all the 10 digits used exactly once.

The temptation now would be to see what would happen if we divided these two reversal sequences of the numbers from 1 − 9, that is,

$$\frac{987654321}{123456789}$$
$$= 8.00000007290000066339000060368490549353263999911470239\ldots$$

All we can say here to the audience is that it is almost equal to 8. By the way, a curious feature of this number is that the number of 0s in the first group is 7, then the next group of 0s has 5, then 3, and then 1. And so you have another pattern with which to entertain folks!

Now, just to go "full circle" and take the original number 987654321 and interchange the last 2 digits to get 987654312 and divide that by 8 (the "approximate answer from previous division), look at the surprising result we arrive at: $\frac{987654312}{8} = 123456789$. Patterns in mathematics seem to always appear — that's what makes the subject so beautiful!

Here are a few more such strange calculations — this time using multiplication — where on either side of the equal sign all nine digits are represented exactly once: $291,548,736 = 8 \times 92 \times 531 \times 746$, and also for $124,367,958 = 627 \times 198,354 = 9 \times 26 \times 531,487$.

Another example of a calculation, where all the digits are used exactly once (not counting the exponent), is $567^2 = 321,489$. This also works for the following: $854^2 = 729,316$. These are, apparently, the only two squares that result in a number, which allows all the digits to be represented once.

Here is one for the audience to ponder: What is the smallest square number that is composed of all 9 digits used exactly once excluding 0? The answer is: $11,826^2 = 139,854,276$. It is then expected that the question will arise: what is the largest square number that is composed of all 9 digits used exactly once excluding 0? Here the answer is: $30,384^2 = 923,187,456$.

If we now include the 0, the smallest square number that is composed of all 10 digits used exactly once is: $32,043^2 = 1,026,753,849$, and the largest square number that is composed of all 10 digits used exactly once is $99,066^2 = 9,814,072,356$. A motivated audience might look for larger numbers whose square produces numbers that include all 10 digits used perhaps more than once. In any case, you should be well fortified to entertain your audience with the above amazing relationships.

While we are on the topic of keeping the same digits used in a calculation, consider the subtraction examples where we subtract two number with like digits only to find that the difference uses only these digits being represented in the subtraction. You will impress (and entertain) your audience by telling them that there are only five such examples, and they are:

$$1980 - 0891 = 1089$$
$$2961 - 1692 = 1269$$
$$3870 - 0783 = 3087$$
$$5823 - 3285 = 2538$$
$$9108 - 8019 = 1089$$

Notice that, in this last case, the numbers being subtracted are reversals of one another and they lead to a very unusual number 1089. This number will fascinate us a bit further in this chapter (page 102).

There seems to be no end to our journey of using all 9 digits in an arithmetic setting. Here is another one. When we take the square and the cube of the number 69, we get two numbers that together use all the ten digits exactly once: $69^2 = 4,761$ and $69^3 = 328,509$. That is, the two numbers 4761 and 328,509 together represent all ten digits. Such amusing examples which can be easily presented with the help of a calculator further exhibit the hidden beauty of mathematics.

## Favorable Numerical Arrangements of Digits 1–9

When wanting to entertain a group with a rather simple task, ask them to arrange the numbers 1, 2, 3, 4, 5, 6, 7, 8, 9 in sequence using only addition and subtraction to reach the number 100.

Here is one possible solution: $123 - 45 - 67 + 89 = 100$.

We are offering some other solutions that you might get from your audience just so you will be properly prepared.

$$123 + 4 - 5 - 67 - 89 = 100$$
$$123 + 45 - 67 + 8 - 9 = 100$$
$$123 - 4 - 5 - 6 - 7 + 8 - 9 = 100$$
$$12 - 3 - 4 + 5 - 6 + 7 + 89 = 100$$
$$12 + 3 + 4 + 5 - 6 - 7 + 89 = 100$$
$$1 + 23 - 4 + 5 + 6 + 78 - 9 = 100$$
$$1 + 2 + 34 - 5 + 67 - 8 + 9 = 100$$
$$12 + 3 - 4 + 5 + 67 + 8 + 9 = 100$$
$$1 + 23 - 4 + 56 + 7 + 8 + 9 = 100$$
$$1 + 2 + 3 - 4 + 5 + 6 + 78 + 9 = 100$$
$$-1 + 2 - 3 + 4 + 5 + 6 + 78 + 9 = 100$$

If this is not enough of an entertainment exercise for the group, you might have them try to do this in reverse, such as: $9 + 8 + 76 + 5 - 4 + 3 + 2 + 1 = 100$. Here are several more to properly fortify you as you entertain the group with the challenge of creating sums of 100 using the numbers in reverse order.

$$98 - 76 + 54 + 3 + 21 = 100$$
$$9 - 8 + 76 + 54 - 32 + 1 = 100$$
$$98 - 7 - 6 - 5 - 4 + 3 + 21 = 100$$
$$9 - 8 + 7 + 65 - 4 + 32 - 1 = 100$$
$$9 - 8 + 76 - 5 + 4 + 3 + 21 = 100$$
$$98 - 7 + 6 + 5 + 4 - 3 - 2 - 1 = 100$$
$$98 + 7 - 6 + 5 - 4 + 3 - 2 - 1 = 100$$
$$98 + 7 + 6 - 5 - 4 - 3 + 2 - 1 = 100$$
$$98 + 7 - 6 + 5 - 4 - 3 + 2 + 1 = 100$$
$$98 - 7 + 6 + 5 - 4 + 3 - 2 + 1 = 100$$
$$98 - 7 + 6 - 5 + 4 + 3 + 2 - 1 = 100$$
$$98 + 7 - 6 - 5 + 4 + 3 - 2 + 1 = 100$$
$$98 - 7 - 6 + 5 + 4 + 3 + 2 + 1 = 100$$
$$9 + 8 + 76 + 5 + 4 - 3 + 2 - 1 = 100$$
$$-9 + 8 + 76 + 5 - 4 + 3 + 21 = 100$$
$$-9 + 8 + 7 + 65 - 4 + 32 + 1 = 100$$
$$-9 - 8 + 76 - 5 + 43 + 2 + 1 = 100$$

You might also extend that challenge by asking the group to create the number 100 using all 10 digits using the operations of addition and multiplication. Here we are asked to create 100 by using all digits from 0 to 9. One such answer is $(9 \times 8) + 7 + 6 + 5 + 4 + 3 + 2 + 1$.

## Using All Nine Digits

We offer here a rather difficult challenge but one that could be quite entertaining and that is to find a way of representing the number 100 using all nine digits in the form of a mixed-number fraction. There are

11 ways to do that and they are by no means simple, but perhaps after seeing a few of them your audience may discover other ways to accomplish this feat. Here are the 11 possibilities:

$$3\frac{69258}{714}, \quad 81\frac{5643}{297}, \quad 81\frac{7524}{396}, \quad 82\frac{3546}{197}, \quad 91\frac{5742}{638}, \quad 91\frac{5823}{647},$$

$$91\frac{7524}{836}, \quad 94\frac{1578}{263}, \quad 96\frac{1428}{357}, \quad 96\frac{1752}{438}, \quad 96\frac{2148}{537}$$

## More Fun with All Ten Digits

You now might further challenge the group to arrange the digits 0–9 by arranging them in fraction form to reach the number 1. Here is one possible solution: $\frac{35}{70} + \frac{148}{296} = \frac{1}{2} + \frac{1}{2} = 1$.

Another challenge that could be entertaining is for members of your group to create the number 10 using all 10 digits. Here they are asked to create 10 by using all digits from 0 to 9. One answer is $1\frac{35}{70} + 8\frac{46}{92} = 10$.

## Four Numbers Make 100

A more challenging question can be led from simple illustrations, where you might feel that there is a simple solution and yet get frustrated trying to find that solution. Supposing you are asked to create the number 100 using exactly four 5s. Here, the solution is rather simple $(5 + 5) \times (5 + 5) = 100$. We can also challenge our audience to use four 9s to make the number 100. This would be a bit more challenging, as it would appear as $99\frac{9}{9}$. And now comes a real challenge to see if they can do that using four 7s. After you have let them struggle for a while you might expose one solution: $\frac{7}{.7} \times \frac{7}{.7} = \frac{7}{\frac{7}{10}} \times \frac{7}{\frac{7}{10}} = \frac{70}{7} \times \frac{70}{7} = 10 \times 10 = 100$. At this point you might have them discover other ways of using four of the same digit to make 100.

# Creating Number Out of Fours

While we are on the task of creating numbers in a creative fashion, you might challenge your audience to see how many numbers they can create using only the number 4 four times. Here we offer a beginning for this task.

$$1 = \frac{4+4}{4+4} = \frac{\sqrt{44}}{\sqrt{44}}$$

$$2 = \frac{4 \cdot 4}{4+4} = \frac{4-4}{4} + \sqrt{4}$$

$$3 = \frac{4+4+4}{4} = \sqrt{4} + \sqrt{4} - \frac{4}{4}$$

$$4 = \frac{4-4}{4} + 4 = \frac{\sqrt{4 \cdot 4 \cdot 4}}{4}$$

$$5 = \frac{4 \cdot 4 + 4}{4}$$

$$6 = \frac{4+4}{4} + 4 = \frac{4\sqrt{4}}{4} + 4$$

$$7 = \frac{44}{4} - 4 = \sqrt{4} + 4 + \frac{4}{4}$$

$$8 = 4 \cdot 4 - 4 - 4 = \frac{4(4+4)}{4}$$

$$9 = \frac{44}{4} - \sqrt{4} = 4\sqrt{4} + \frac{4}{4}$$

$$10 = 4 + 4 + 4 - \sqrt{4}$$

$$11 = \frac{4}{4} + \frac{4}{.4}$$

$$12 = \frac{4 \cdot 4}{\sqrt{4}} + 4 = 4 \cdot 4 - \sqrt{4} - \sqrt{4}$$

$$13 = \frac{44}{4} + \sqrt{4}$$

$$14 = 4 \cdot 4 - 4 + \sqrt{4} = 4 + 4 + 4 + \sqrt{4}$$

$$15 = \frac{44}{4} + 4 = \frac{\sqrt{4} + \sqrt{4} + \sqrt{4}}{4}$$

$$16 = 4 \cdot 4 - 4 + 4 = \frac{4 \cdot 4 \cdot 4}{4}$$

$$17 = 4 \cdot 4 + \frac{4}{4}$$

$$18 = \frac{44}{\sqrt{4}} - 4 = 4 \cdot 4 + 4 - \sqrt{4}$$

$$19 \frac{4 + \sqrt{4}}{.4} + 4$$

$$20 = 4 \cdot 4 + \sqrt{4} + \sqrt{4}$$

$$21 = 4! - 4 + \frac{4}{4}$$

$$22 = 4 \cdot 4 + 4 + \sqrt{4} = \frac{4}{4}(4!) - \sqrt{4}$$

$$23 = 4! - \sqrt{4} + \frac{4}{4} = 4! - 4^{4-4}$$

$$24 = 4 \cdot 4 + 4 + 4$$

$$25 = \left(4 + \frac{4}{4}\right)^{\sqrt{4}}$$

$$26 = 4! + \sqrt{(4+4-4)} = \frac{4}{4}(4!) + \sqrt{4}$$

$$27 = 4! + 4 - \frac{4}{4}$$

$$28 = 4! + \sqrt{4} \cdot 4 - 4 = 44 - 4 \cdot 4 = (4+4) \cdot 4 - 4$$

$$29 = 4! + 4 + \frac{4}{4}$$

$$30 = 4! + 4 + 4 - \sqrt{4}$$

## Multiplying by 12345679 by a Multiple of 3

Here we can have some fun by multiplying the number 12,345,679 (notice the 8 is missing) by various multiples of 3 to get some surprising results. Take for example:

$$12345679 \times 45 = 555,555,555$$
$$12345679 \times 48 = 592,592,592$$

Here are some more examples:

$$12345679 \times 63 = 777,777,777$$
$$12345679 \times 54 = 666,666,666$$

There are many other surprising results through these multiple-of-3 multiplications. Each will probably bring surprise and pleasure.

## A Surprising Division

It is sometimes interesting to entertain people who have a calculator at hand and would like to see a pretty result, namely, where our 10 digits keep repeating as shown below.

$\frac{137,174,210}{1,111,111,111} = 0.$**1234567890**12345678901**234567890**, this beautiful result requires no further explanation.

# Products of 91 and the Numbers 1–9

In order to appreciate the next suggested multiplications, we need to do all the multiplications from 1 to 9, and then admire the results vertically.

$$91 \times 1 = 091$$
$$91 \times 2 = 182$$
$$91 \times 3 = 273$$
$$91 \times 4 = 364$$
$$91 \times 5 = 455$$
$$91 \times 6 = 546$$
$$91 \times 7 = 637$$
$$91 \times 8 = 728$$
$$91 \times 9 = 819$$

# Some Number Peculiarities

Number oddities need not necessarily be restricted to a single number. There are times when these oddities appear with partner numbers. Consider the addition of the two numbers: $192 + 384 = 576$. Your audience may ask, what is so special about this addition? Have them look at the outside digits (bold): **192** + **384** = **576**. They are in numerical sequence left to right (1, 2, 3, 4, 5, 6) and then reversing to get the rest of the 9 digits (7, 8, 9). They might have also noticed that the three numbers that we used in this addition example also have a strange relationship as you can see from the following:

$$192 = 1 \cdot 192$$
$$384 = 2 \cdot 192$$
$$576 = 3 \cdot 192$$

## Understanding Division

With our continuous reliance on the calculator many folks don't think twice about division. There are times when it is useful, and perhaps also entertaining, to provoke an audience with a simple arithmetic conundrum. For example, ask your audience what is the smallest number which is divisible by 13 and at the same time when divided by any of the numbers from 2 to 12, inclusive, will leave a remainder of 1. At first reckoning, the audience may be a bit perplexed. However, little thought will begin a trial and error approach.

It would be wise for you to be prepared to explain how this can be easily approached with just a little bit of algebra. The smallest number which can be divided by each of the numbers from 2 to 12 and leave no remainder is the product $12 \times 11 \times 5 \times 7 \times 3 \times 2 = 27{,}720$. Therefore, the number we seek is $27{,}720n + 1$, which when divided by 13 gives

$$\frac{27{,}720n+1}{13} = 2{,}132n + \frac{4n+1}{13}$$

For this to be divisible by 13, the number $4n + 1$ must be a multiple of 13, which is clearly the smallest number for $n = 3$. Thus, the number we seek is $27{,}720n + 1$ for $n = 3$, and we get $27{,}720 \times 3 + 1 = 83{,}161$.

At first sight, the solution to this conundrum may seem a bit complicated, but a slow and thoughtful presentation can make it rather entertaining as well as instructive.

## A Hidden Curiosity in the Sequence of Initial Cubic Numbers

There are times when a seemingly "harmless" arrangement of numbers produces an unexpected result. This sort of surprise can be entertaining and enlightening. Let's consider the first eight perfect cubes: $1^3, 2^3, 3^3, 4^3, 5^3, 6^3, 7^3, 8^3, \ldots$, whose values are:

1, 8, 27, 64, 125, 216, 343, 512. When we take the differences between these cubes, we get 7, 19, 37, 61, 91, 127, 169, which at first sight do not appear to have any special property. However, when we take the differences of these numbers, we get 12, 18, 24, 30, 36, 42, whose common difference is 6. This gives some further meaning to the list of consecutive cubes.

Audiences can often be impressed with unusual number relationships. Of course, as a presenter, you would need to indicate the rarity of these relationships and then, of course, they can verify them with a calculator to make sure that they are correct. However, once established, they can truly appreciate it. Here are a few of such relationships, where the following equation is common among them: $d^{2n} + e^{2n} + f^{2n} = a^{n} + b^{n} + c^{n}$,

$$3^4 + 4^4 + 5^4 = 5^2 + 19^2 + 24^2$$
$$3^8 + 4^8 + 5^8 = 5^4 + 19^4 + 24^4$$

Similarly, for the relationship $a^{n} + b^{n} + c^{n} = d^{n} + e^{n} + f^{n}$

$$1^1 + 6^1 + 8^1 = 2^1 + 4^1 + 9^1$$
$$1^2 + 6^2 + 8^2 = 2^2 + 4^2 + 9^2$$
$$7^2 + 34^2 + 41^2 = 14^2 + 29^2 + 43^2$$
$$7^4 + 34^4 + 41^4 = 14^4 + 29^4 + 43^4$$
$$1^1 + 5^1 + 8^1 + 12^1 = 2^1 + 3^1 + 10^1 + 11^1$$
$$1^2 + 5^2 + 8^2 + 12^2 = 2^2 + 3^2 + 10^2 + 11^2$$
$$1^3 + 5^3 + 8^3 + 12^3 = 2^3 + 3^3 + 10^3 + 11^3$$

These are not the only such relationships that share this common pattern. It might be a nice challenge to seek others with more than 4 numbers on each side of equal sign. It is the unusualness of these relationships that makes them special and worthy of awe by an audience.

## An Unusual Number

The number 76923, when multiplied by some other numbers, such as 2, 7, 5, 11, 6 and 8, provides another striking pattern. It would be interesting to see if the patterns are immediately notice by your audience.

$$76923 \times 2 = \mathbf{153846}$$
$$76923 \times 7 = \mathbf{538461}$$
$$76923 \times 5 = \mathbf{384615}$$
$$76923 \times 11 = \mathbf{846153}$$
$$76923 \times 6 = \mathbf{461538}$$
$$76923 \times 8 = \mathbf{615384}$$

We notice that the first digits of these products exhibit the same number as the first product, namely 153846. Here, as well the diagonal from the upper right to the lower left of the products is represented by 6, and lines parallel to the diagonal are also represented by the same digits.

## Products of 37 and Multiples of 3

While we are having fun discovering unusual multiplication results, consider multiplying 37 by these numbers: 3, 6, 9, 12, 15, 18, 21, 24, 27, and we find that the respective products are: 111, 222, 333, 444, 555, 666, 777, 888, 999. Once again, to avoid distractions, a calculator would be very useful.

## Products of 3367 and Multiples of 33

Here are some more entertaining products to consider with the calculator at hand.

$$3367 \times 33 = 111,111$$
$$3367 \times 66 = 222,222$$
$$3367 \times 99 = 333,333$$
$$3367 \times 132 = 444,444$$
$$3367 \times 165 = 555,555$$
$$\vdots$$
$$3367 \times 297 = 999,999$$

## A Double Number Curiosity

We will consider here a rather entertaining phenomenon of double numbers such as beginning with the number 418. If we write the number 418,418, which for our purposes we will call a double number, we can get back to the single number 418, by dividing 418,418 consecutively by 7, 11, and 13 as shown below:

$$418,418 \div 7 = 59,774$$

$$59,774 \div 11 = 5,434$$

$$5,434 \div 13 = 418$$

Your audience may want to know why this happens. Have them consider the product $7 \times 11 \times 13 = 1001$, which when multiplied by a three-digit number will always give you the double number. In other words, we multiply $618 \times 1001 = 618,618$.

A motivated audience might ask if this can be done for larger numbers, say double numbers of 8 digits in length, such as 23,562,356. A clever participant might suggest dividing this number by 2356, which would result in $23,562,356 \div 2356 = 10,001$. This time would need to find the factors of 10,001, which are 73 and 137. Therefore, we take a four-digit number, such as 1836 and multiplied by 73 to get 134,028, and then multiply that by 137, we get 18,361,836, which is the sought-after double number.

If someone asks, how do we get a double number when we start off with a five-digit number, the answer would be to multiply by 100,001, which would require to multiplications once by 11, and once by 9091.

Suppose someone asks how we can get a three-digit number transformed into a quadruple number, namely where the number repeats the 3 digits 4 times. This would be a bit more complicated and would require a more sophisticated calculator, because this time we would need to multiply the given three-digit number by 1,001,001,001, which would require 5 multiplications of the original three-digit number by 7, 11, 13, 101, and 9,901.

You can entertain your audience further along using this technique for creating double and other multiple numbers.

## Strange Relationships

You can surely impress an audience with certain number pairs that yield the same product even when both numbers are reversed, for example, if $12 \times 42 = 504$. When we reverse each of the two numbers, we get $21 \times 24 = 504$. The same is true for the number pair: 36 and 84, since $36 \times 84 = 3024 = 63 \times 48$.

At this point your audience may wonder, if this will happen with any pair of numbers. The answer is that it will only work with the following 14 pairs of numbers:

| | |
|---|---|
| $12 \times 42 = 21 \times 24 = 504$ | $12 \times 63 = 21 \times 36 = 756$ |
| $12 \times 84 = 21 \times 48 = 1008$ | $13 \times 62 = 31 \times 26 = 806$ |
| $13 \times 93 = 31 \times 39 = 1209$ | $14 \times 82 = 41 \times 28 = 1148$ |
| $23 \times 64 = 32 \times 46 = 1472$ | $23 \times 96 = 32 \times 69 = 2208$ |
| $24 \times 63 = 42 \times 36 = 1512$ | $24 \times 84 = 42 \times 48 = 2016$ |
| $26 \times 93 = 62 \times 39 = 2418$ | $34 \times 86 = 43 \times 68 = 2924$ |
| $36 \times 84 = 63 \times 48 = 3024$ | $46 \times 96 = 64 \times 69 = 4416$ |

A careful inspection of these 14 pairs of numbers will reveal that in each case the product of the tens digits of each pair of numbers is equal to the product of the units digits. For the curious audience, we can easily justify this algebraically as follows:

For the numbers $k_1, k_2, k_3,$ and $k_4$, we have

$$k_1 \cdot k_2 = (10a + b) \cdot (10c + d) = 100ac + 10ad + 10bc + bd, \text{ and}$$
$$k_3 \cdot k_4 = (10b + a) \cdot (10d + c) = 100bd + 10bc + 10ad + ac.$$

Here $a, b, c, d$ represent any of the ten digits: 0, 1, 2, ... , 9, where $a \neq 0$ and $c \neq 0$.

We would like to show that $k_1 \cdot k_2 = k_3 \cdot k_4$. Therefore, $100ac + 10ad + 10bc + bd = 100bd + 10bc + 10ad + ac$, then $100ac + bd = 100bd + ac$, and $99ac = 99bd$, or $ac = bd$, which is what we observed earlier.

## Some Numerology Gags

This is not seen by some as mathematics, but we can digress and revel in some numerology — where words would be replaced by numbers and still make sense. Take, for example, the following word sentence: SIX + SIX + SIX = NINE + NINE.

Our task is to find numerals that can replace the letters and still be correct. There are two such:

$942 + 942 + 942 = 1413 + 1413$, where the number 4 replaces the I and all the other letters are represented by unique digits.

Another example is: $472 + 472 + 472 = 0708 + 0708$, where the number 7 has replaced the I, and the other letters are represented by unique digits. You need not ask folks to search for other replacement numbers, since these are the only ones that exist for this word equation.

## The Digits Remain in Use

You can entertain folks with simple multiplication, of course, having a calculator handy makes the experience much more fluid. Here are several multiplication examples that continue to use only the digits of the two given numbers in the product:

$$30 \times 51 = 1530$$
$$21 \times 87 = 1827$$
$$80 \times 86 = 6880$$
$$21 \times 81 = 2187$$
$$60 \times 21 = 1260$$
$$93 \times 15 = 1395$$
$$41 \times 35 = 1435$$

Naturally, this can be taken for larger numbers as well. If you really want to impress your audience, take a pair of larger numbers, and consider the following: $9{,}162{,}361{,}086 \times 1{,}234{,}554{,}321 = 11{,}311{,}432{,}469{,}283{,}552{,}606$. Checking this would be quite some task, but it would broaden the picture a bit for the audience, especially those who choose to investigate this further.

## More Strange Relationships

Earlier when we were marveling over the number 9, we have seen the relationship that

$$9 + 9 = 18 \text{ and } 9 \times 9 = 81.$$

Now you can entertain your audience with several more similar relationships, such as:

$$24 + 3 = 27 \quad \text{and} \quad 24 \times 3 = 72$$
$$47 + 2 = 49 \quad \text{and} \quad 47 \times 2 = 94$$
$$497 + 2 = 499 \quad \text{and} \quad 497 \times 2 = 994$$

Showing these to an audience typically draws a surprise as to the beauty of mathematics that seems all too often hidden.

Here is a quick amusement: Which number is 4 times its reversal? The number is 2178, which is $2178 \times 4 = 8712$. Are there others? An ambitious reader may want to explore this.

Another quickie is the following: $2^7 - 5^3 = 5 - 2$, or $13^3 - 3^3 = 13 - 3$, and there are more curiosities with which to entertain folks.

## Curious Relationships

We can also consider the curiosity of upside-down numbers. This is truly an opportunity for you to entertain your audience with these 2 equal sums: $69 + 98 + 86 = 96 + 68 + 89 = 253$.

The sum of squares of these flipped numbers are also equal: $69^2 + 98^2 + 86^2 = 96^2 + 68^2 + 89^2 = 21,761$. This rather unique relationship should probably get applause from the audience.

If that isn't enough, we offer still another rather curious relationship. Consider the following equality, where from left to right every pair of numbers are reflections of one another: $1181 + 1811 + 8188 + 8818 = 1118 + 1888 + 8111 + 8881 = 19,998$. What makes this so special is that when we take the squares of each of these numbers we still maintain an equality: $1181^2 + 1811^2 + 8188^2 + 8818^2 = 1118^2 + 1888^2 + 8111^2 + 8881^2 = 149,494,950$, and amazingly the sums of the cubes are also equal: $1181^3 + 1811^3 + 8188^3 + 8818^3 = 1118^3 + 1888^3 + 8111^3 + 8881^3 = 1,242,200,007,576$. And if you feel we cannot top this, then consider that we flip these numbers around, upside down, or in mirror image, and all of the above will still be true. This will surely overwhelm your audience with wonder!

## Kaprekar Numbers

There are other numbers that have somewhat similar unusual peculiarities. Sometimes these peculiarities can be understood and justified through an algebraic representation, while at other times a peculiarity is simply a quirk of the base-10 number system. In any case, these numbers provide some rather entertaining amusements that ought to motivate us to look for other such peculiarities or oddities.

Consider, for example, the number 297. When we take the square of that number, we get $297^2 = 88,209$, which, if we were to split it up, strangely enough, the sum of the two numbers of the split results in the original number: $88 + 209 = 297$. Such a number is called a *Kaprekar number*, named after the Indian mathematician Dattatreya Ramchandra Kaprekar (1905–1986) who discovered numbers with this characteristic. Here are a few more examples of Kaprekar numbers.

$9^2 = 81...$                  $8 + 1 = 9$

$45^2 = 2025$    ...         $20 + 25 = 45$

$$55^2 = 3025 \quad \ldots \qquad 30 + 25 = 55$$
$$703^2 = 494{,}209 \quad \ldots \qquad 494 + 209 = 703$$
$$2{,}728^2 = 7{,}441{,}984 \quad \ldots \qquad 744 + 1{,}984 = 2728$$
$$4{,}879^2 = 23{,}804{,}641 \quad \ldots \qquad 238 + 04{,}641 = 4879$$
$$142{,}857^2 = 20408122449 \qquad 20{,}408 + 122{,}449 = 142{,}857$$

A more extensive listing of the Kaprekar numbers is provided for the ambitious audience who may be motivated to seek other such numbers.

Table of Kaprekar Numbers.

| Kaprekar number | Square of the number | | Decomposition |
|---|---|---|---|
| 1 | $1^2 =$ | 1 | $1 = 1$ |
| 9 | $9^2 =$ | 81 | $8 + 1 = 9$ |
| 45 | $45^2 =$ | 2025 | $20 + 25 = 45$ |
| 55 | $55^2 =$ | 3025 | $30 + 25 = 55$ |
| 99 | $99^2 =$ | 9801 | $98 + 01 = 99$ |
| 297 | $297^2 =$ | 88,209 | $88 + 209 = 297$ |
| 703 | $703^2 =$ | 494,209 | $494 + 209 = 703$ |
| 999 | $999^2 =$ | 998,001 | $998 + 001 = 999$ |
| 2223 | $2223^2 =$ | 4,941,729 | $494 + 1729 = 2223$ |
| 2728 | $2728^2 =$ | 7,441,984 | $744 + 1984 = 2728$ |
| 4879 | $4879^2 =$ | 23,804,641 | $238 + 4641 = 4879$ |
| 4950 | $4950^2 =$ | 24,502,500 | $2450 + 2500 = 4950$ |
| 5050 | $5050^2 =$ | 25,502,500 | $2550 + 2500 = 5050$ |
| 5292 | $5292^2 =$ | 28,005,264 | $28 + 5264 = 5292$ |
| 7272 | $7272^2 =$ | 52,881,984 | $5288 + 1984 = 7272$ |
| 7777 | $7777^2 =$ | 60,481,729 | $6048 + 1729 = 7777$ |
| 9999 | $999^2 =$ | 99,980,001 | $9998 + 1 = 9999$ |
| 17,344 | $17{,}344^2 =$ | 300,814,336 | $3008 + 14{,}336 = 17{,}344$ |
| 22,222 | $22{,}222^2 =$ | 493,817,284 | $4938 + 17{,}284 = 22{,}222$ |
| 38,962 | $38{,}962^2 =$ | 1,518,037,444 | $1518 + 37{,}444 = 38{,}962$ |
| 77,778 | $77{,}778^2 =$ | 6,049,417,284 | $60{,}494 + 17{,}284 = 77{,}778$ |

(*Continued*)

(*Continued*)

| | | |
|---|---|---|
| 82,656 | $82,656^2 =$ 6,832,014,336 | $68,320 + 14,336 = 82,656$ |
| 95,121 | $95,121^2 =$ 9,048,004,641 | $90,480 + 04,641 = 95,121$ |
| 99,999 | $99,999^2 =$ 9,999,800,001 | $99,998 + 00,001 = 99,999$ |
| 142,857 | $142,857^2 =$ 20,408,122,449 | $20,408 + 122,449 = 142,857$ |
| 148,149 | $148,149^2 =$ 21,948,126,201 | $21,948 + 126,201 = 148,149$ |
| 181,819 | $181,819^2 =$ 33,058,148,761 | $33,058 + 148,761 = 181,819$ |
| 187,110 | $187,110^2 =$ 35,010,152,100 | $35,010 + 152,100 = 187,110$ |

The next larger Kaprekar numbers are 208495, 318682, 329967, 351352, 356643, 390313, 461539, 466830, 499500, 500500, 533170, 857143, .... .

There are also further variations, such as the number 45, which we would consider a *Kaprekar triple*, since it behaves as follows: $45^3 = 91,125 = 9 + 11 + 25 = 45$. Other Kaprekar triples are 1, 8, 10, 297, and 2322. Curiously enough the number 297, which we previously demonstrated as a Kaprekar number, is also a Kaprekar triple, since $297^3 = 26,198,073$ and $26 + 198 + 073 = 297$. A motivated audience will normally seek to find other Kaprekar triples.

## An Unusual Arithmetic Sequence

There are times when an unsuspecting and rather unusual sequence of numbers produces some sparkling results, take, for example, the sequence:

15,873, 31,746, 47,619, 63,492, 79,365, 95,238, 111,111, 126,984, 142,857, where the common difference between the numbers is 15,873. When each of these numbers is multiplied by 7, we amazingly get the following products, respectively: 111,111, 222,222, 333,333, 444,444, 555,555, 666,666, 777,777, 888,888, 999,999.

Certainly, the seventh number in the sequence when multiplied by 7 will not produce a great surprise, but the others are quite astonishing and will probably be awed by the audience.

# Triangular Numbers

It should be well known among your friends that square numbers such as 1, 4, 9, 16, 25, etc., when representing a number of points, can be arranged in the shape of a square. Analogously, one can do quite a bit of entertaining with numbers that can represent points, which can be arranged in the shape of an equilateral triangle and which are, therefore, called *triangular numbers*. We show the first few triangular numbers as they can be arranged in the form of an equilateral triangle in Figure 1.14.

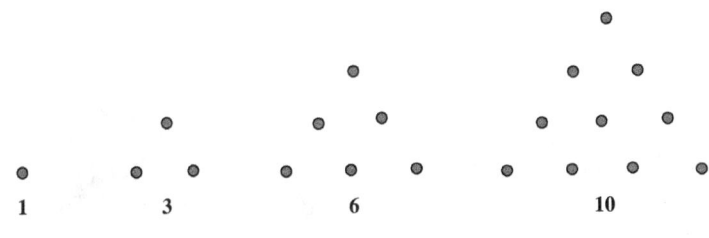

**Figure 1.14**

To make the reader properly fortified, here are the triangular numbers, which are less than 10,000:

1, 3, 6, 10, 15, 21, 28, 36, 45, 55, 66, 78, 91, 105, 120, 136, 153, 171, 190, 210, 231, 253, 276, 300, 325, 351, 378, 406, 435, 465, 496, 528, 561, 595, 630, 666, 703, 741, 780, 820, 861, 903, 946, 990, 1035, 1081, 1128, 1176, 1225, 1275, 1326, 1378, 1431, 1485, 1540, 1596, 1653, 1711, 1770, 1830, 1891, 1953, 2016, 2080, 2145, 2211, 2278, 2346, 2415, 2485, 2556, 2628, 2701, 2775, 2850, 2926, 3003, 3081, 3160, 3240, 3321, 3403, 3486, 3570, 3655, 3741, 3828, 3916, 4005, 4095, 4186, 4278, 4371, 4465, 4560, 4656, 4753, 4851, 4950, 5050, 5151, 5253, 5356, 5460, 5565, 5671, 5778, 5886, 5995, 6105, 6216, 6328, 6441, 6555, 6670, 6786, 6903, 7021, 7140, 7260, 7381, 7503, 7626, 7750, 7875, 8001, 8128, 8256, 8385, 8515, 8646, 8778, 8911, 9045, 9180, 9316, 9453, 9591, 9730, 9870.

Perhaps the most easily detectable property of triangular numbers (we will represent the $n$th triangular number as $T_n$) is that they each are the sum of the first $n$ consecutive natural numbers. Although your audience can experiment to justify this statement, it might be part of your entertaining act to show them the first few triangular numbers:

$$T_1 = 1$$
$$T_2 = 1 + 2 = 3$$
$$T_3 = 1 + 2 + 3 = 6$$
$$T_4 = 1 + 2 + 3 + 4 = 10$$
$$T_5 = 1 + 2 + 3 + 4 + 5 = 15$$
$$T_6 = 1 + 2 + 3 + 4 + 5 + 6 = 21$$
$$T_7 = 1 + 2 + 3 + 4 + 5 + 6 + 7 = 28$$

However, since these triangular numbers result from an arithmetic series, the formula to find the $n$th triangular number (which was probably presented in the high school curriculum) can be found with the following formula: $T_n = \frac{n(n+1)}{2}$.

As a small digression, it is clever to point out at this time that the sum of initial consecutive *odd* integers will always equal a square number:

$$1$$
$$1 + 3 = 4$$
$$1 + 3 + 5 = 9$$
$$1 + 3 + 5 + 7 = 16$$
$$1 + 3 + 5 + 7 + 9 = 25$$
$$1 + 3 + 5 + 7 + 9 + 11 = 36$$

This is just the beginning of where the properties of triangular numbers take off, and where you can almost limitlessly entertain your friends. Let us now enjoy some of these truly unexpected properties of triangular numbers.

1. First, let's see how the triangular number can relate to square numbers. The sum of any two consecutive triangular numbers is equal to a square number, for example,

$$T_1 + T_2 = 1 + 3 = 4 = 2^2$$
$$T_5 + T_6 = 15 + 21 = 36 = 6^2$$
$$T_6 + T_7 = 21 + 28 = 49 = 7^7$$

2. By inspecting the above list of triangular numbers, you will notice that a triangular number never appears to end with a units digit of 2, 4, 7, or 9. This is true throughout all of the triangular numbers.

3. The number 3 is the only triangular number that is also a prime number.

4. All perfect numbers (which are numbers equal to the sum of their proper divisors, such as 6, 28, 496, 8128, etc.) are also triangular numbers.

5. If 1 is added to 9 times a triangular number, the result will be another triangular number, such as in the case of the following example: $9 \cdot T_3 + 1 = 9 \cdot 6 + 1 = 54 + 1 = 55$, which is the tenth triangular number. You might want to try this with other triangular numbers to see that this truly works. We shall do that here as well: $9 \cdot T_5 + 1 = 9 \cdot 15 + 1 = 136$, which is the 16th triangular number.

6. We can take the previous curiosity and expand it to consider what would happen if we multiplied a triangular number by 8 instead of by 9 and then add 1. We find that if 1 is added to 8 times a triangular number, the result will be a square number, as in the case for $8 \cdot T_3 + 1 = 8 \cdot 6 + 1 = 48 + 1 = 49 = 7^2$. Or we can try this for the seventh triangular number $T_7$ to get: $8 \cdot T_7 + 1 = 8 \cdot 28 + 1 = 225 = 15^2$. Once again, you can entertain your audience with this unexpected connection of triangular numbers with square numbers.

7. The sum of $n$ consecutive cubes beginning with 1 equals the square of the $n$th triangular number. That is, $T_n^2 = 1^3 + 2^3 + 3^3 + 4^3 + \cdots + n^3$. As an example, consider the following sum of the first five

cubes: $T_5^2 = 1^3 + 2^3 + 3^3 + 4^3 + 5^3 = 1 + 8 + 27 + 64 + 125 = 225 = 15^2$ — once again there is a connection between square numbers and triangular numbers.

8. Relating triangular numbers to our previously encountered palindromic numbers, we find that some triangular numbers are also palindromic numbers — reading the same forward and backward. The first few of these are: 1, 3, 6, 55, 66, 171, 595, 666, 3003, 5995, 8778, 15051, 66066, 617716, 828828, 1269621, 1680861, 3544453, 5073705, 5676765, 6295926, 351335153, 61477416, 178727871, 1264114621, 1634004361, etc.

9. Once again, relating triangular numbers and square numbers, we find that there are infinitely many triangular numbers, which are also square numbers. The first few of these "square-triangular" are: $1 = 1^2$, $36 = 6^2$, $1225 = 35^2$, $41616 = 204^2$, $1413721 = 1189^2$, $48024900 = 6930^2$, $1631432881 = 40391^2$, etc.

   Naturally, we can generate these square-triangular numbers with the following formula: $Q_n = 34Q_{n-1} - Q_{n-2} + 2$, where $Q_n$ represents the $n$th square-triangular number.

   An interesting peculiarity among these square-triangular numbers is that all even square-triangular numbers are a multiple of 9.

10. To add some more fun to triangular numbers we should take note that there are some triangular numbers, which, when the order of the digits is reversed, also produce a triangular number. The first few of these are: 1, 3, 6, 10, 55, 66, 120, 153, 171, 190, 300, 351, 595, 630, 666, 820, 3003, 5995, 8778, 15051, 17578, 66066, 87571, 156520, 180300, 185745, 547581, etc.

11. Another curiosity about triangular numbers is that certain members of the set of triangular numbers can be paired so that the sum and the difference of these pairs also result in a triangular number. Here are a few examples of these triangular-number pairs: (15, 21) yields: 21 − 15 = 6, and 21 + 15 = 36, and both 6 and 36 are also triangular numbers; (105, 171) yields: 171 − 105 = 66, and 105 + 171 = 276, where both 66 and 276 are also triangular numbers.

12. To further fascinate your audience with the peculiarities of triangular numbers, have them consider that among the practically boundless properties of triangular numbers there are only six triangular numbers that can be represented as the product of 3 consecutive numbers. These are $T_3 = 6 = 1 \cdot 2 \cdot 3$, $T_{15} = 120 = 4 \cdot 5 \cdot 6$, $T_{20} = 210 = 9 \cdot 10 \cdot 11$, $T_{44} = 990 = 9 \cdot 10 \cdot 11$, $T_{608} = 185136 = 56 \cdot 57 \cdot 58$, $T_{22736} = 258474216 = 636 \cdot 637 \cdot 638$. Of these, the number $T_{15} = 120$ is particularly "gifted" in that it can be represented as the product of three, four and five consecutive numbers. There has not been another such triangular number found that has this property. That is, $T_{15} = 120 = 4 \cdot 5 \cdot 6 = 2 \cdot 3 \cdot 4 \cdot 5 = 1 \cdot 2 \cdot 3 \cdot 4 \cdot 5$. In this last case, we have sought merely consecutive numbers (rather than consecutive triangular numbers as we have earlier), we can show that there are some triangular numbers that are the product of 2 consecutive numbers, as the following examples: $T_3 = 6 = 2 \cdot 3$, $T_{20} = 210 = 14 \cdot 15$, $T_{119} = 7140 = 84 \cdot 85$, and $T_{696} = 242556 = 492 \cdot 493$.

13. Another oddity to entertain folks with is to indicate that there are only six triangular numbers that are comprised of one type of digit. They are 1, 3, 6, 55, 66, 666.

14. Among the Fibonacci numbers (we have already introduced the Fibonacci numbers on pages 20 and 27) there are only four known triangular numbers: 1, 3, 21, and 55.

15. To write a positive integer as the sum of triangular numbers, you would never need more than three triangular numbers. For example, the first ten positive integers can be represented by the sum of triangular numbers as follows: **1**, **2** $= 1 + 1$, **3**, **4** $= 1 + 3$, **5** $= 1 + 1 + 3$, **6**, **7** $= 1 + 6$, **8** $= 1 + 1 + 6$, **9** $= 3 + 3 + 3$, **10** $= 1 + 3 + 6$, etc. This can then be tried with some of the triangular numbers shown earlier to see if the audience can meet the challenge of finding the proper sums.

16. Here is another exercise that could be entertaining, namely, to find enough examples to show that very fourth power of an integer greater than 1 can be expressed as the sum of two triangular numbers:

$$2^4 = 16 = 6 + 10 = 1 + 15 = T_3 + T_4 = T_1 + T_5$$
$$3^4 = 81 = T_8 + T_9 = T_5 + T_{11}$$

$$4^4 = 256 = T_{15} + T_{16} = T_{11} + T_{19}$$
$$5^4 = 625 = T_{24} + T_{25} = T_{19} + T_{29}$$
$$6^4 = 1296 = T_{35} + T_{36} = T_{29} + T_{41}$$
$$7^4 = 2401 = T_{48} + T_{49} = T_{41} + T_{55}$$

17. More fun with triangular numbers involves showing that the sum of successive powers of 9 will result in a triangular number, as we can see from the following first few such examples:

$$9^0 = T_1$$
$$9^0 + 9^1 = T_4$$
$$9^0 + 9^1 + 9^2 = T_{13}$$
$$9^0 + 9^1 + 9^2 + 9^3 = T_{40}$$
$$9^0 + 9^1 + 9^2 + 9^3 + 9^4 = T_{121}$$

18. Our last relationship on triangular numbers that can surely entertain the audience, which is now more familiar with triangular numbers, is one that can complement the first one. Notice the pattern of the sums in the following.

$$T_1 + T_2 + T_3 = T_4$$
$$T_5 + T_6 + T_7 + T_8 = T_9 + T_{10}$$
$$T_{11} + T_{12} + T_{13} + T_{14} + T_{15} = T_{16} + T_{17} + T_{18}$$
$$T_{19} + T_{20} + T_{21} + T_{22} + T_{23} + T_{24} = T_{25} + T_{26} + T_{27} + T_{28}$$
$$T_{29} + T_{30} + T_{31} + T_{32} + T_{33} + T_{34} + T_{35} = T_{36} + T_{37} + T_{38} + T_{39} + T_{40}$$

By this point, you ought to have enough material to entertain folks with the surprising relationships that involve triangular numbers. The introduction of triangular numbers, which seems to be missing from most school curricula, offers a delightful insight into the beauty of mathematics — one that seems to hold many opportunities for further exploration.

# A Mathematical Conjecture

Much of the mathematics is taught in the schools is somehow justified in some kind of logical argument or proof. There are a number of phenomena in mathematics that appear to be true, but have never been substantiated, or proved. These are referred to as mathematical conjectures. In some cases, computers have been able to generate enormous numbers of examples to support a statement's veracity, but that (amazingly) does not make it true for all cases. To conclude the truth of a statement, we must have a legitimate proof that it is true for *all* cases!

Perhaps one of the most famous mathematical conjectures that has frustrated mathematicians for centuries was presented by the German mathematician Christian Goldbach (1690–1764), to the famous Swiss mathematician Leonhard Euler (1707–1783), in a June 7, 1742 letter, where he posed the following statement, which to this day has yet to be *proved* true. Commonly known as *Goldbach's Conjecture*, it states the following:

*Every even number greater than 2 can be expressed as the sum of two prime numbers.*

We can begin with the following list of even numbers and their prime-number sums and then continue it to somehow convince ourselves that it continues on — apparently — indefinitely.

| Even numbers greater than 2 | Sum of two prime numbers |
|:---:|:---:|
| 4 | 2 + 2 |
| 6 | 3 + 3 |
| 8 | 3 + 5 |
| 10 | 3 + 7 |
| 12 | 5 + 7 |

*(Continued)*

(*Continued*)

| | |
|---|---|
| 14 | 7 + 7 |
| 16 | 5 + 11 |
| 18 | 7 + 11 |
| 20 | 7 + 13 |
| ⋮ | ⋮ |
| 48 | 19 + 29 |
| ⋮ | ⋮ |
| 100 | 3 + 97 |

There have been substantial attempts by famous mathematicians throughout the centuries to try to prove, or somehow justify, this conjecture. In 1855, A. Desboves verified Goldbach's Conjecture for up to 10,000 places. Yet, in 1894, the famous German mathematician Georg Cantor (1845–1918) (regressing a bit) showed that the conjecture was true for all even numbers up to 1000. It was then shown by N. Pipping in 1940 to be true for all even numbers up to 100,000. By 1964, with the aid of a computer, it was extended to 33,000,000, and in 1965 this was extended to 100,000,000; and later in 1980 to 200,000,000. Then, in 1998, the German mathematician Jörg Richstein showed that Goldbach's conjecture was true for all even numbers up to 400 trillion. On February 16, 2008, Oliveira e Silva extended this to 1.1 quintillion! (i.e., $1.1 \times 10^{18} = 1,100,000,000,000,000,000$). Prize money of $1,000,000 has been offered for a proof of this conjecture. To date, this has not been claimed, as the conjecture has never been proved for all cases.

Goldbach also had a Second Conjecture, which is as follows:

*Every odd number greater than 5 can be expressed as the sum of three primes.*

Again, we shall present you with a few examples and let you continue the list as far as you wish.

| Odd numbers greater than 5 | Sum of three prime numbers |
|:---:|:---:|
| 7 | $2 + 2 + 3$ |
| 9 | $3 + 3 + 3$ |
| 11 | $3 + 3 + 5$ |
| 13 | $3 + 5 + 5$ |
| 15 | $5 + 5 + 5$ |
| 17 | $5 + 5 + 7$ |
| 19 | $5 + 7 + 7$ |
| 21 | $7 + 7 + 7$ |
| $\vdots$ | $\vdots$ |
| 51 | $3 + 17 + 31$ |
| $\vdots$ | $\vdots$ |
| 77 | $5 + 5 + 67$ |
| $\vdots$ | $\vdots$ |
| 101 | $5 + 7 + 89$ |

These unsolved problems have tantalized many mathematicians over the centuries, and although no proof has yet been found, more evidence (built with the help of computers) suggests that these must be true, and additionally no counterexample has been found. Interestingly, the efforts to try to prove this conjecture have led to some significant discoveries in mathematics that might have gone hidden without this impetus. They provoke us as well as provide sources of entertainment.

## The Pattern of Powers

When we take powers of numbers, say the numbers from 1 to 20, we notice that there is a definite pattern of the units digits of these larger numbers and the chart in Figure 1.15 shows the units digit of the numbers considered in the left-hand column when they are squared, cubed, taken to the 4th power, taken to the 5th power, and so on. In the column of the squared numbers there is a symmetry **within** the first 10 numbers and then each succeeding 10 numbers follows that symmetry. The cubes of these numbers are arranged in groups of 10, that is the first 10 and the second 10 have the same units digits. The 4th powers have a symmetry in each interval of 10 numbers and comparing the first 10 to the second 10 cubes there is again a similarity. The most "comfortable" pattern is seen in the 5th powers and the 9th powers. Yet, there are similarities among and between various powers when it comes to further inspecting the units digits. This should be entertaining provided the audience does not have to do too many calculations.

| Number | Squared | Cubed | 4th | 5th | 6th | 7th | 8th | 9th |
|--------|---------|-------|-----|-----|-----|-----|-----|-----|
| 1 | 1 | 1 | 1 | 1 | 1 | 1 | 1 | 1 |
| 2 | 4 | 8 | 6 | 2 | 4 | 8 | 6 | 2 |
| 3 | 9 | 7 | 1 | 3 | 9 | 7 | 1 | 3 |
| 4 | 6 | 4 | 6 | 4 | 6 | 4 | 6 | 4 |
| 5 | 5 | 5 | 5 | 5 | 5 | 5 | 5 | 5 |
| 6 | 6 | 6 | 6 | 6 | 6 | 6 | 6 | 6 |
| 7 | 9 | 3 | 1 | 7 | 9 | 3 | 1 | 7 |
| 8 | 4 | 2 | 6 | 8 | 4 | 2 | 6 | 8 |
| 9 | 1 | 9 | 1 | 9 | 1 | 9 | 1 | 9 |
| 10 | 0 | 0 | 0 | 0 | 0 | 0 | 0 | 0 |
| 11 | 1 | 1 | 1 | 1 | 1 | 1 | 1 | 1 |
| 12 | 4 | 8 | 6 | 2 | 4 | 8 | 6 | 2 |
| 13 | 9 | 7 | 1 | 3 | 9 | 7 | 1 | 3 |
| 14 | 6 | 4 | 6 | 4 | 6 | 4 | 6 | 4 |
| 15 | 5 | 5 | 5 | 5 | 5 | 5 | 5 | 5 |
| 16 | 6 | 6 | 6 | 6 | 6 | 6 | 6 | 6 |
| 17 | 9 | 3 | 1 | 7 | 9 | 3 | 1 | 7 |
| 18 | 4 | 2 | 6 | 8 | 4 | 2 | 6 | 8 |
| 19 | 1 | 9 | 1 | 9 | 1 | 9 | 1 | 9 |
| 20 | 0 | 0 | 0 | 0 | 0 | 0 | 0 | 0 |

**Figure 1.15**

# Howlers

In the early years of schooling, we learned to reduce fractions to make them more manageable, and we were taught various ways to do it correctly. Suppose you are asked to reduce the fraction $\frac{26}{65}$, and do it in the following way: $\frac{2\cancel{6}}{\cancel{6}5} = \frac{2}{5}$.

That is, just cancel out the 6's to get the right answer. Is this procedure correct? Can it be extended to other fractions? If this is so, then we were surely treated unfairly by our elementary school teachers, who made us do much more work to reduce fractions to lowest terms. Let's look at what was done here and see if it can be generalized. In his 1959 book, *Fallacies in Mathematics*,[4] the British mathematician Edwin A. Maxwell (1907–1987) refers to the following cancellations as "howlers."

$$\frac{1\cancel{6}}{\cancel{6}4} = \frac{1}{4}, \quad \frac{2\cancel{6}}{\cancel{6}5} = \frac{2}{5}$$

Perhaps when someone did the fraction reductions this way, and still got the right answer, it could just make you howl. As we look at this awkward — yet easy — procedure, we could actually use it to reduce the following fractions to lowest terms: $\frac{16}{64}, \frac{19}{95}, \frac{26}{65}, \frac{49}{98}$.

After you have reduced each of the fractions to lowest terms in the usual manner, one may ask why it couldn't have been done in the following way.

$$\frac{1\cancel{6}}{\cancel{6}4} = \frac{1}{4}$$

$$\frac{1\cancel{9}}{\cancel{9}5} = \frac{1}{5}$$

$$\frac{2\cancel{6}}{\cancel{6}5} = \frac{2}{5}$$

$$\frac{4\cancel{9}}{\cancel{9}8} = \frac{4}{8} = \frac{1}{2}$$

At this point, the folks to whom you show this may be somewhat amazed. Their first reaction is probably to ask, if this can be done to any fraction composed of two-digit numbers of this sort? Can you find another fraction (comprised of two-digit numbers) where this type of cancellation will work? You might cite $\frac{5\cancel{5}}{\cancel{5}5} = \frac{5}{5} = 1$ as an illustration of this type of cancellation. This will, clearly, hold true for all two-digit multiples of 11.

The justification for this weird technique can be shown to those people with a good working knowledge of elementary algebra, which can "explain" this awkward occurrence. Are the four fractions above the *only* ones (composed of different two-digit numbers) where this type of cancellation will hold true? We will come back to this question shortly.

Consider the fraction

$$\frac{10x+a}{10a+y}$$

The above four cancellations were such that when canceling the $a$'s the fraction was equal to $\frac{x}{y}$.

Therefore,

$$\frac{10x+a}{10a+y}=\frac{x}{y}$$

This yields:

$$y(10x+a)=x(10a+y)$$
$$10xy+ay=10ax+xy$$
$$9xy+ay=10ax$$

and so, $y=\frac{10ax}{9x+a}$.

At this point, we shall inspect this equation. It is necessary that $x$, $y$ and $a$ are integers, since they were digits in the numerator and denominator of a fraction. It is now our task to find the values of $a$ and $x$ for which $y$ will also be integral. To avoid a lot of algebraic manipulation, we will set up a chart which will generate values of $y$ from the equation $y=\frac{10ax}{9x+a}$. Remember that $x$, $y$ and $a$ must be single-digit integers. We have in Figure 1.16 a portion of the table that shows the values of $y=\frac{10ax}{9x+a}$. (Notice that the cases where $x=a$ are excluded.)

| x\a | 1 | 2 | 3 | 4 | 5 | 6 | ... | 9 |
|---|---|---|---|---|---|---|---|---|
| 1 |  | $\frac{20}{11}$ | $\frac{30}{12}$ | $\frac{40}{13}$ | $\frac{50}{14}$ | $\frac{60}{15}=4$ |  | $\frac{90}{18}=5$ |
| 2 | $\frac{20}{19}$ |  | $\frac{60}{21}$ | $\frac{80}{22}$ | $\frac{100}{23}$ | $\frac{120}{24}=5$ |  |  |
| 3 | $\frac{30}{28}$ | $\frac{60}{29}$ |  | $\frac{120}{31}$ | $\frac{150}{32}$ | $\frac{180}{33}$ |  |  |
| 4 |  |  |  |  |  |  |  | $\frac{360}{45}=8$ |
| ⋮ |  |  |  |  |  |  |  |  |
| 9 |  |  |  |  |  |  |  |  |

**Figure 1.16**

The portion of the chart in Figure 1.16 generated the 4 integral values of $y$; that is, when $x = 1$, $a = 6$ then $y = 4$ and when $x = 2$, $a = 6$ then $y = 5$. These values yield the fractions $\frac{16}{64}$ and $\frac{26}{65}$, respectively, and when $x = 1$ and $a = 9$, yielding $y = 5$, also when $x = 4$ and $a = 9$, yielding $y = 8$. These give us the fractions $\frac{19}{95}$ and $\frac{49}{98}$, respectively. This should justify that there are only 4 such fractions composed of two-digit numbers.

You may now wonder, if there are fractions composed of numerators and denominators of more than 2 digits, where this strange type of cancellation holds true. Try this type of cancellation with $\frac{499}{998}$. You should find that $\frac{499}{998} = \frac{4}{8} = \frac{1}{2}$.

A pattern is now emerging, and you may realize that the following can also qualify under this scheme:

$$\frac{49}{98} = \frac{499}{998} = \frac{4999}{9998} = \frac{49999}{99998} = \frac{4}{8} = \frac{1}{2}$$

$$\frac{16}{64} = \frac{166}{664} = \frac{1666}{6664} = \frac{16666}{66664} = \frac{1}{4}$$

$$\frac{19}{95} = \frac{199}{995} = \frac{1999}{9995} = \frac{19999}{99995} = \frac{1}{5}$$

$$\frac{2\cancel{6}}{\cancel{6}5} = \frac{2\cancel{66}}{\cancel{66}5} = \frac{2\cancel{666}}{\cancel{666}5} = \frac{2\cancel{6666}}{\cancel{6666}5} = \frac{2}{5}$$

Enthusiastic audience members may wish to justify these extensions of the original howlers. For those who, at this point, have a further desire to seek out additional fractions, which permit this strange cancellation, they should consider the following fractions and then verify the legitimacy of this strange cancellation. If they have been properly motivated, they may then set out to discover more such fractions.

$$\frac{3\cancel{3}2}{8\cancel{3}0} = \frac{32}{80} = \frac{2}{5}$$

$$\frac{3\cancel{8}5}{8\cancel{8}0} = \frac{35}{80} = \frac{7}{16}$$

$$\frac{1\cancel{3}8}{\cancel{3}45} = \frac{18}{45} = \frac{2}{5}$$

$$\frac{2\cancel{7}5}{7\cancel{7}0} = \frac{25}{70} = \frac{5}{14}$$

$$\frac{1\cancel{6}\cancel{3}}{\cancel{3}2\cancel{6}} = \frac{1}{2}$$

Aside from providing an algebraic justification for this strange method, we offer here some more of these "howlers."

$$\frac{4\cancel{8}4}{\cancel{8}47} = \frac{4}{7} \quad \frac{\cancel{5}45}{65\cancel{5}} = \frac{5}{6} \quad \frac{\cancel{4}24}{7\cancel{4}2} = \frac{4}{7} \quad \frac{24\cancel{9}}{\cancel{9}96} = \frac{24}{96} = \frac{1}{4}$$

$$\frac{4\cancel{8}4\cancel{8}4}{\cancel{8}4\cancel{8}47} = \frac{4}{7} \quad \frac{\cancel{5}4\cancel{5}45}{65\cancel{5}4\cancel{5}4} = \frac{5}{6} \quad \frac{\cancel{4}2\cancel{4}24}{7\cancel{4}2\cancel{4}2} = \frac{4}{7}$$

$$\frac{\cancel{3}2\cancel{4}3}{43\cancel{2}\cancel{4}} = \frac{3}{4} \quad \frac{6\cancel{4}86}{8\cancel{6}4\cancel{8}} = \frac{6}{8} = \frac{3}{4}$$

$$\frac{14714}{71468} = \frac{14}{68} = \frac{7}{34} \qquad \frac{878048}{987804} = \frac{8}{9}$$

$$\frac{1428571}{4285713} = \frac{1}{3} \qquad \frac{2857142}{8571426} = \frac{2}{6} = \frac{1}{3} \qquad \frac{3461538}{4615384} = \frac{3}{4}$$

$$\frac{767123287}{876712328} = \frac{7}{8} \qquad \frac{3243243243}{4324324324} = \frac{3}{4}$$

$$\frac{1025641}{4102564} = \frac{1}{4} \qquad \frac{3243243}{4324324} = \frac{3}{4} \qquad \frac{4571428}{5714285} = \frac{4}{5}$$

$$\frac{4848484}{8484847} = \frac{4}{7} \qquad \frac{5952380}{9523808} = \frac{5}{8} \qquad \frac{4285714}{6428571} = \frac{4}{6} = \frac{2}{3}$$

$$\frac{5485485}{6548548} = \frac{5}{6} \qquad \frac{6923076}{9230768} = \frac{6}{8} = \frac{3}{4} \qquad \frac{4242424}{7424242} = \frac{4}{7}$$

$$\frac{5384615}{7538461} = \frac{5}{7} \qquad \frac{2051282}{8205128} = \frac{2}{8} = \frac{1}{4} \qquad \frac{3116883}{8311688} = \frac{3}{8}$$

$$\frac{6486486}{8648648} = \frac{6}{8} = \frac{3}{4} \qquad \frac{484848484}{848484847} = \frac{4}{7}$$

This peculiarity shows how elementary algebra can be used to investigate a number theory situation, and fortunately for us, one that is also quite amusing.

## Typographical Errors That *Are* Correct

Just for entertainment, here are some typographical errors that turn out to be correct; some examples where the × was missing and misplaced, such as with:

$$73 \times 9 \times 42 = 7 \times 3942$$
$$73 \times 9 \times 420 = 7 \times 39420$$

Then there are some examples where the × and the exponents were dropped out, such as with:

$$2^5 \times \frac{25}{31} = 25\frac{25}{31}$$

$$2^5 \times 9^2 = 2592$$

$$3^4 \times 425 = 34425$$

$$3^4 \times 4250 = 344250$$

$$11^2 \times 9\frac{1}{3} = 1129\frac{1}{3}$$

$$21^2 \times 4\frac{9}{11} = 2124\frac{9}{11}$$

$$13^2 \times 7\frac{6}{7} = 1327\frac{6}{7}$$

There are also some very complicated versions of these curious typographical errors, such as

$$13^2 \times 7857142\frac{6}{7} = 1327857142\frac{6}{7}$$

There are many more such strange coincidences where symbols are deleted or misread and still the correct answer results. We provide this merely as an entertaining activity!

## Wrong Arithmetic

Imagine an elementary school student who is learning multiplication of fractions and finds that the following appears to be correct: $\frac{1}{4} \times \frac{8}{5} = \frac{18}{45} = \frac{2}{5}$. In other words, the student feels that to do the multiplication you need merely to combine the digits in the numerator and the denominator to get the right answer, since clearly $\frac{1}{4} \times \frac{8}{5} = \frac{8}{20} = \frac{2}{5}$.

Unconvinced with your reasoning that this doesn't work, the student shows you another example, where this method does work, such as $\frac{2}{6} \times \frac{6}{5} = \frac{26}{65} = \frac{2}{5}$. Does this imply that the student has come up with a new method of multiplying fractions? This will certainly give your audience something to think about. The student may very well say, we can flip the two fractions and it still works, as we see with $\frac{6}{2} \times \frac{5}{6} = \frac{65}{26} = \frac{5}{2}$, which is a correct result with a wrong procedure. There are 14 such examples, where it works so that you can demonstrate the limits of this weird and incorrect multiplication. They are as follows:

$$\frac{1}{4} \times \frac{8}{5} = \frac{18}{45}, \quad \frac{1}{2} \times \frac{5}{4} = \frac{15}{24}, \quad \frac{1}{6} \times \frac{4}{3} = \frac{14}{63}, \quad \frac{1}{6} \times \frac{6}{4} = \frac{16}{64}, \quad \frac{1}{9} \times \frac{9}{5} = \frac{19}{95},$$

$$\frac{4}{9} \times \frac{9}{8} = \frac{49}{98}, \quad \frac{2}{6} \times \frac{6}{5} = \frac{26}{65}$$

where each of these can be flipped to get another 7 such examples. Using simple algebra, we can show that these are the only 7 examples, where the digits are not the same. You would let $a$, $b$, $c$ and $d$ be the digits from 1 – 9, so that $\frac{a}{b} \times \frac{c}{d} = \frac{10a+c}{10b+d}$, which then can be reduced to the equation $ac(10b+d) = bd(10a+c)$, which then leads you to the 7 examples above and their flips.

## The Ever-Present Number 6174

There are some numbers in our decimal system that have unique characteristics. One such number is 6147. To exhibit this strange characteristic, we begin by selecting any four-digit number, where the digits are not all the same. Following the procedure that we will provide below, everyone, using any such four-digit number, will end up with a number 6174. To make matters simpler and less distracting, a calculator will be useful here.

1. *Begin by selecting any four-digit number — except one that has all digits the same.*

2.  *Rearrange the digits of the number so that they form the largest number possible. (In other words, write the number with the digits in descending order).*

3.  *Then rearrange the digits of the number so that they form the smallest number possible. (That is, write the number with the digits in ascending order. Zeros can take the first few places).*

4.  *Subtract these two numbers (obviously, the smaller from the larger).*

5.  *Take this difference and continue the process, over and over and over, until you notice something disturbing happening. Don't give up before something unusual happens.*

Eventually all the members of your group to whom you will show this entertaining exercise will eventually arrive at the number **6174** — perhaps after one subtraction, or after several subtractions. Once they arrive at this number 6174, they will find themselves in an endless loop, which means that by continuing the process with the number 6174, they will continue to end up with 6174. Remember that everyone involved will have begun with an arbitrarily selected number. Although some readers might be motivated to investigate this further, others will just sit back in awe.

Here is an example of how this works with our arbitrarily selected starting number 3927.

The *largest* number formed with these digits is 9732
The *smallest* number formed with these digits is 2379
The difference is 7353

Now using this number, 7353, we continue the process:

The largest number formed with these digits is 7533
The smallest number formed with these digits is 3357
The difference is 4176

Again, we repeat the process.

The largest number formed with these digits is 7641
The smallest number formed with these digits is 1467
The difference is 6174

When one arrives at 6174, the number continuously reappears. Notice that the largest number that can be formed with these digits 7641 and the smallest number is 1467 and the difference as we have seen above is 6174. Remember, all this began with an *arbitrarily selected* **four-digit** number and will always end up with the number 6174, which then gets you into an endless loop by continuously getting back to 6174. The unusualness of this number will clearly entertain your audience.

This nifty loop was first discovered by the Indian mathematician, Dattatreya Ramchandra Kaprekar in 1946.[5] We often refer to the number 6174 as the *Kaprekar constant*.

By the way, just as an aside, the number 6174 is also divisible by the sum of its digits. That is, $\frac{6174}{6+1+7+4} = \frac{6174}{18} = 343$.

## Some Variations of the Kaprekar Constants:

- If you choose a two-digit number (not one with two same digits), then the Kaprekar constant would be 81 and you would end up in a loop of length 5: [81, 63, 27, 45, 09 (,81)].
  There is no loop of length 1 for two-digit numbers, as we had earlier.
- If you choose a three-digit number (not one with all same digits), then the Kaprekar constant would be 495 and you would end up in a loop of length 1: [495 (, 495)].
- If you choose a four-digit number (not one with all same digits), then the Kaprekar constant would be 6174 — as we have seen before — and you end up with a loop of length 1: [**6174 (, 6174)**].
- If you choose a five-digit number (not one of all same digits), then there are three Kaprekar constants: 53,955, 61,974, and 62,964.

One of length 2: [53 955, 59 994, 53 955)]
and two of length 4: [61 974, 82 962, 75 933, 63 954 (, 61 974)]
[62 964, 71 973, 83 952, 74 943 (, 62 964)]

You can follow this scheme with six-digit numbers and you will also find yourself getting into a loop. One number you may find leading you into the loop is 840,852, but do not let this stop you from further investigating this mathematical curiosity.[6] For example, consider the digit-sum of each difference. Since the sum of the digits of the subtrahend and the minuend[7] are the same, the difference will have a digit sum that is a multiple of 9. For three- and four-digit numbers, the digit-sum is 18. In the case of five- and six-digit numbers, the digit-sum appears to be 27. It follows that, for seven- and eight-digit numbers, the digit-sum is 36. Yes, you will find that the digit-sum, when this technique is used on nine- and ten-digit number, is 45. You will be pleasantly surprised when you check to see what the digit sum is for even larger numbers.

## The Special Number 1089

There are other numbers that also lend themselves very nicely to exhibiting a unique situation. One such involves a number 1089. You can surely entertain an audience by showing them the following procedure which will always end up with a number 1089 regardless of which numbers one begins with. Supposing we select any randomly selected three-digit number, where the units and hundreds digit are not the same. We then follow this step-by-step procedure and then admire the unsuspecting outcome.

> *Choose any three-digit number (where the unit and hundreds digit are not the same).*
> Suppose we arbitrarily select the number **835**.
> *Reverse the digits of the number you have selected.*
> We then get the number **538**.
> *Subtract the two numbers (naturally, the larger minus the smaller).*
> Our difference is: 835 − 538 = **297**.

*Once again, reverse the digits of this difference.*
Giving us the number: **792**.
*Now, add you last two numbers.*
We then add the last two numbers to get: $297 + 792 = $ **1089**.

When you try this with a group of people, each of whom has chosen a different initial three-digit number, each member of the group should have gotten the same result as ours, namely 1089, If some individuals claim they got a different number, then clearly they made a calculation error! In any case, your group will probably be astonished that, regardless of which number was initially selected, they all got the same result as we did, 1089.

If the original three-digit number had the same units and hundreds digit, then we would get a 0 after the first subtraction, such as for a starting number of 373, since $373 - 373 = 0$; this would ruin our model. Before reading on, convince yourself that this scheme will work for other numbers. How does it happen? Is this a "freak property" of this number? Did we do something devious in our calculations?

This illustration of a mathematical oddity depends on the operations. We assumed that any number we chose would lead us to 1089. How can we be sure? Well, we could try all possible three-digit numbers to see if it works. That would be tedious and not particularly elegant. An investigation of this oddity requires nothing more than some knowledge of elementary algebra. Yet, were we to try to test all possibilities, it would be interesting to determine to how many such three-digit numbers we would have to apply to this scheme. Remember, we can only use those three-digit numbers whose units digit and hundreds digit are not the same.

For the reader who might be curious about this phenomenon, we will provide an algebraic explanation as to why it "works." We shall represent the arbitrarily selected three-digit number, ***htu*** as $100h + 10t + u$, where $h$ represents the hundreds digit, $t$ represents the tens digit, and $u$ represents the units digit.

Let $h > u$, which would be the case either in the number you selected or the reverse of it.

In the subtraction, $u-h<0$; therefore, take 1 from the tens place (of the minuend) making the units place $10+u$. Since the tens digits of the 2 numbers to be subtracted are equal, and 1 was taken from the tens digit of the minuend, then the value of this digit is $10(t-1)$. The hundreds digit of the minuend is $h-1$, because 1 was taken away to enable subtraction in the tens place, making the value of the tens digit $10(t-1)+100=10(t+9)$.

We can now do the first subtraction:

$$
\begin{array}{lll}
100(h-1) & +100(t+9) & +(u+10) \\
100u & +10t & +h \\
\hline
100(h-u-1) & +10(9) & +u-h+10
\end{array}
$$

Reversing the digits of this difference gives us:

$$100(u-h+10)+10(9)+(h-u-1)$$

Now adding these last two expressions gives us:

$$100(9)+10(18)+(10-1)=\underline{1089}$$

It is important to stress that algebra enables us to inspect the arithmetic process, regardless of the number used in this entertaining curiosity.

Before we leave the number 1089, we should point out to the enthusiastic audience that is enchanted with these surprises that this number still harbors another oddity, namely, $33^2=1089=65^2-56^2$, which is unique among two-digit numbers. By this time, you must agree that there is a particular beauty in the number **1089**. Are you hooked? If you're not yet convinced about the peculiarity of the number 1089, read on.

The number 1089 also lends itself to another interesting numerical pattern. When we multiplied 1089 by each of the digits from 1 to 9 the products have a very curious property: the units digits decreased by 1 each time, beginning with the number 9; the tens digits decreased by 1 each time beginning with the number 8; the hundreds digits increase each time starting with 0; and the thousandths digits increased each times again with the number 1.

| | |
|---|---|
| 1 × 1089 | 1089 |
| 2 × 1089 | 2178 |
| 3 × 1089 | 3267 |
| 4 × 1089 | 4356 |
| 5 × 1089 | 5445 |
| 6 × 1089 | 6534 |
| 7 × 1089 | 7623 |
| 8 × 1089 | 8712 |
| 9 × 1089 | 9801 |

**Figure 1.17**

Also, of note is that the last entry in Figure 1.17 shows that 9801 is a multiple of its reversal 1089. You might want to further entertain your audience by telling them that there is only one number of five distinct digits, whose *multiple* is a reversal of the original number. That number is 21,978, since 4 × 21,978 = 87,912 is its reverse number. The number 1089 gives rise to similarly structured numbers that provide even more fodder for entertaining an audience.

Remember that 1089 × 9 = 9801, which is the reversal of the original number. The same property holds for 10989 × 9 = 98901, and similarly, 109989 × 9 = 989901. You should recognize that we altered the original number, 1089, by inserting a 9 in the middle of the number to get 10989, and extended that by inserting 99 in the middle of the number 1089 to get 109989. It would be nice to conclude from this that each of the following numbers has the same property: 1099989, 10999989, 109999989, 1099999989, 10999999989, and so on.

As a matter of fact, there is only one other number with four or fewer digits, where a multiple of itself is equal to its reversal, and that is the number 2178 (which just happens to be 2 × 1089), since 2178 × 4 = 8712. Wouldn't it be nice if we could extend this as we did with the above example by inserting 9s into the middle of the number to generate other numbers that have the same property? Yes, the following is true:

$$21978 \times 4 = 87912$$
$$219978 \times 4 = 879912$$
$$2199978 \times 4 = 8799912$$
$$21999978 \times 4 = 87999912$$
$$219999978 \times 4 = 879999912$$
$$2199999978 \times 4 = 8799999912$$
$$\vdots$$

## The Irrepressible Number 10989 and Others

The previous enchantment can be extended to larger numbers. There are many oddities in mathematics that can provide entertainment as well as motivate research. We will consider one here. This phenomenon will leave the reader to discover why this oddity works; however, it will also provide the reader with a cute little peculiarity to share with friends. For example, suppose you ask your friends to write any four-digit number on a piece of paper, where the difference between the first and last digits is greater than 1. Then have them interchange the first and last digits. Next, they are to subtract smaller from the larger of these two numbers. Once again, interchange the first and last digits of this resulting subtraction, and then added two the previously obtained numbers. They should have gotten a number 10989.

To illustrate how this works, we shall select a random four-digit number, were the units digit and the thousandths digit differ by more than 1. Suppose we select the number 5367. Following the previously described procedure, we will interchange the first and last digits to get the number 7365. We then subtract these two numbers 7365 − 5367 = 1998. Once again, we interchange of first and last digits to get 8991, and then add this to the previous-obtained number to get 1998 + 8991 = 10989, as predicted!

Let's try this process with the five-digit number 97356. When we interchange the first and last digits, we get the number 67359. We now subtract these two numbers: $97356 - 67359 = 29997$. We then interchange the first and last digits to get 79992. Now adding the last two numbers: $29997 + 79992 = 109989$. This will always be our final result, regardless of which five-digit number we start with.

If you want to further impress your friends, have them select a six-digit number and follow the same procedure. They will always find that the end result will be 1099989. If we are starting with a seven-digit number, the result will analogously be 10999989. This, then, continues on with the same pattern for larger numbers. This experience can generate some genuine interest in the nature of numbers and of course make mathematics come alive.

## The Ulam–Collatz Loop

There are times when we think of the beauty in nature as magical. Is nature magical? Some feel that when something is truly surprising and "neat," it is beautiful. From that standpoint, we will show a seemingly "magical" property in mathematics. This is one that has baffled mathematicians for many years and still no one knows why it happens. Try it, you'll like it.

We begin by asking you to follow two rules as you work with any *arbitrarily selected* number.

*If the number is odd, then multiply by 3 and add 1.*
*If the number is even, then divide by 2.*

Regardless of the number you select, after continued repetition of the process, you will always end up with the number 1. Once again something for your audience to marvel at, especially when they realize that everyone began with a different number.

Let's try it for the *arbitrarily selected* number **7**:
7 is odd, therefore, multiply by 3 and add 1 to get: $7 \cdot 3 + 1 = \mathbf{22}$.

22 is even, so we simply divide by 2 to get **11**.
11 is odd, so we multiply by 3 and add 1 to get **34**.
34 is even, so we divide by 2 to get **17**.
17 is odd, so we multiply by 3 and add 1 to get **52**.
52 is even, so we divide by 2 to get **26**.
26 is even, so we divide by 2 to get **13**.
13 is odd, so we multiply by 3 and add 1 to get **40**.
40 is even, so we divide by 2 to get **20**.
20 is even, therefore, divide by 2 to get **10**.
10 is even, therefore, divide by 2 to get **5**.
5 is odd, so we multiply by 3 and add 1 to get **16**.
16 is even, so we divide by 2 to get **8**.
8 is even, so we divide by 2 to get **4**.
4 is even, so we divide by 2 to get **2**.
2 is also even, so we again divide by 2 to get **1**.

If we were to continue, we would find ourselves in a loop.

As you can see with the following continuation: 1 is odd, so we multiply by 3 and add 1 to get **4**, is even, so we divide by 2 to get **2**. 2 is also even, so we again divide by 2 to get **1,** again! After 16 steps, we end up with a 1, that, if we continue the process, would lead us back again to 4, and then on to the 1 again. We hit a loop!

Therefore, we get the sequence 7, 22, 11, 34, 17, 52, 26, 13, 40, 20, 10, 5, 16, 8, **4, 2, 1** , **4, 2, 1**, ....

To provide you some guidance (and perhaps some encouragement as well), we offer a chart that shows the steps that each number *n* will require in order to reach the 1.

# Ulam-Collatz Sequences for $n = 0, 1, 2, …, 100$

| $n$ | Ulam-Collatz sequence |
|---|---|
| 0 | [0, 0] |
| 1 | [1, 4, 2, 1] = [1, **4**, **2**, **1** (, 4)] |
| 2 | [2, 1, 4, 2] = [2, 1, **4**, **2**, **1** (, 4)] |
| 3 | [3, 10, 5, 16, 8, **4**, **2**, **1** (, 4)] |
| 4 | [**4**, **2**, **1** (, 4)] |
| 5 | [5, 16, 8, **4**, **2**, **1** (, 4)] |
| 6 | [6, 3, 10, 5, 16, 8, **4**, **2**, **1** (, 4)] |
| 7 | [7, 22, 11, 34, 17, 52, 26, 13, 40, 20, 10, 5, 16, 8, **4**, **2**, **1** (, 4)] |
| 8 | [8, **4**, **2**, **1** (, 4)] |
| 9 | [9, 28, 14, 7, 22, 11, 34, 17, 52, 26, 13, 40, 20, 10, 5, 16, 8, **4**, **2**, **1** (, 4)] |
| 10 | [10, 5, 16, 8, **4**, **2**, **1** (, 4)] |
| 11 | [11, 34, 17, 52, 26, 13, 40, 20, 10, 5, 16, 8, **4**, **2**, **1** (, 4)] |
| 12 | [12, 6, 3, 10, 5, 16, 8, **4**, **2**, **1** (, 4)] |
| 13 | [13, 40, 20, 10, 5, 16, 8, 4, 2, 1, 4] |
| 14 | [14, 7, 22, 11, 34, 17, 52, 26, 13, 40, 20, 10, 5, 16, 8, **4**, **2**, **1** (, 4)] |
| 15 | [15, 46, 23, 70, 35, 106, 53, 160, 80, 40, 20, 10, 5, 16, 8, **4**, **2**, **1** (, 4)] |
| 16 | [16, 8, **4**, **2**, **1** (, 4)] |
| 17 | [17, 52, 26, 13, 40, 20, 10, 5, 16, 8, **4**, **2**, **1** (, 4)] |
| 18 | [18, 9, 28, 14, 7, 22, 11, 34, 17, 52, 26, 13, 40, 20, 10, 5, 16, 8, **4**, **2**, **1** (, 4)] |
| 19 | [19, 58, 29, 88, 44, 22, 11, 34, 17, 52, 26, 13, 40, 20, 10, 5, 16, 8, **4**, **2**, **1** (, 4)] |
| 20 | [20, 10, 5, 16, 8, **4**, **2**, **1** (, 4)] |
| 21 | [21, 64, 32, 16, 8, **4**, **2**, **1** (, 4)] |
| 22 | [22, 11, 34, 17, 52, 26, 13, 40, 20, 10, 5, 16, 8, **4**, **2**, **1** (, 4)] |
| 23 | [23, 70, 35, 106, 53, 160, 80, 40, 20, 10, 5, 16, 8, **4**, **2**, **1** (, 4)] |
| 24 | [24, 12, 6, 3, 10, 5, 16, 8, **4**, **2**, **1** (, 4)] |
| 25 | [25, 76, 38, 19, 58, 29, 88, 44, 22, 11, 34, 17, 52, 26, 13, 40, 20, 10, 5, 16, 8, **4**, **2**, **1** (, 4)] |
| 26 | [26, 13, 40, 20, 10, 5, 16, 8, **4**, **2**, **1** (, 4)] |

27 | [27, 82, 41, 124, 62, 31, 94, 47, 142, 71, 214, 107, 322, 161, 484, 242, 121, 364, 182, 91, 274, 137, 412, 206, 103, 310, 155, 466, 233, 700, 350, 175, 526, 263, 790, 395, 1186, 593, 1780, 890, 445, 1336, 668, 334, 167, 502, 251, 754, 377, 1132, 566, 283, 850, 425, 1276, 638, 319, 958, 479, 1438, 719, 2158, 1079, 3238, 1619, 4858, 2429, 7288, 3644, 1822, 911, 2734, 1367, 4102, 2051, 6154, 3077, 9232, 4616, 2308, 1154, 577, 1732, 866, 433, 1300, 650, 325, 976, 488, 244, 122, 61, 184, 92, 46, 23, 70, 35, 106, 53, 160, 80, 40, 20, 10, 5, 16, 8, **4, 2, 1** (, 4)]

28 | [28, 14, 7, 22, 11, 34, 17, 52, 26, 13, 40, 20, 10, 5, 16, 8, **4, 2, 1** (, 4)]

29 | [29, 88, 44, 22, 11, 34, 17, 52, 26, 13, 40, 20, 10, 5, 16, 8, **4, 2, 1** (, 4)]

30 | [30, 15, 46, 23, 70, 35, 106, 53, 160, 80, 40, 20, 10, 5, 16, 8, **4, 2, 1** (, 4)]

31 | [31, 94, 47, 142, 71, 214, 107, 322, 161, 484, 242, 121, 364, 182, 91, 274, 137, 412, 206, 103, 310, 155, 466, 233, 700, 350, 175, 526, 263, 790, 395, 1186, 593, 1780, 890, 445, 1336, 668, 334, 167, 502, 251, 754, 377, 1132, 566, 283, 850, 425, 1276, 638, 319, 958, 479, 1438, 719, 2158, 1079, 3238, 1619, 4858, 2429, 7288, 3644, 1822, 911, 2734, 1367, 4102, 2051, 6154, 3077, 9232, 4616, 2308, 1154, 577, 1732, 866, 433, 1300, 650, 325, 976, 488, 244, 122, 61, 184, 92, 46, 23, 70, 35, 106, 53, 160, 80, 40, 20, 10, 5, 16, 8, **4, 2, 1** (, 4)]

32 | [32, 16, 8, **4, 2, 1** (, 4)]

33 | [33, 100, 50, 25, 76, 38, 19, 58, 29, 88, 44, 22, 11, 34, 17, 52, 26, 13, 40, 20, 10, 5, 16, 8, **4, 2, 1** (, 4)]

34 | [34, 17, 52, 26, 13, 40, 20, 10, 5, 16, 8, **4, 2, 1** (, 4)]

35 | [35, 106, 53, 160, 80, 40, 20, 10, 5, 16, 8, **4, 2, 1** (, 4)]

36 | [36, 18, 9, 28, 14, 7, 22, 11, 34, 17, 52, 26, 13, 40, 20, 10, 5, 16, 8, **4, 2, 1** (, 4)]

37 | [37, 112, 56, 28, 14, 7, 22, 11, 34, 17, 52, 26, 13, 40, 20, 10, 5, 16, 8, **4, 2, 1** (, 4)]

38 | [38, 19, 58, 29, 88, 44, 22, 11, 34, 17, 52, 26, 13, 40, 20, 10, 5, 16, 8, **4, 2, 1** (, 4)]

39 | [39, 118, 59, 178, 89, 268, 134, 67, 202, 101, 304, 152, 76, 38, 19, 58, 29, 88, 44, 22, 11, 34, 17, 52, 26, 13, 40, 20, 10, 5, 16, 8, **4, 2, 1** (, 4)]

40 | [40, 20, 10, 5, 16, 8, **4, 2, 1** (, 4)]

41 | [41, 124, 62, 31, 94, 47, 142, 71, 214, 107, 322, 161, 484, 242, 121, 364, 182, 91, 274, 137, 412, 206, 103, 310, 155, 466, 233, 700, 350, 175, 526, 263, 790, 395, 1186, 593, 1780, 890, 445, 1336, 668, 334, 167, 502, 251, 754, 377, 1132, 566, 283, 850, 425, 1276, 638, 319, 958, 479, 1438, 719, 2158, 1079, 3238, 1619, 4858, 2429, 7288, 3644, 1822, 911, 2734, 1367, 4102, 2051, 6154, 3077, 9232, 4616, 2308, 1154, 577, 1732, 866, 433, 1300, 650, 325, 976, 488, 244, 122, 61, 184, 92, 46, 23, 70, 35, 106, 53, 160, 80, 40, 20, 10, 5, 16, 8, **4, 2, 1** (, 4)]

42 | [42, 21, 64, 32, 16, 8, **4**, **2**, **1** (, 4)]

43 | [43, 130, 65, 196, 98, 49, 148, 74, 37, 112, 56, 28, 14, 7, 22, 11, 34, 17, 52, 26, 13, 40, 20, 10, 5, 16, 8, **4**, **2**, **1** (, 4)]

44 | [44, 22, 11, 34, 17, 52, 26, 13, 40, 20, 10, 5, 16, 8, **4**, **2**, **1** (, 4)]

45 | [45, 136, 68, 34, 17, 52, 26, 13, 40, 20, 10, 5, 16, 8, **4**, **2**, **1** (, 4)]

46 | [46, 23, 70, 35, 106, 53, 160, 80, 40, 20, 10, 5, 16, 8, **4**, **2**, **1** (, 4)]

47 | [47, 142, 71, 214, 107, 322, 161, 484, 242, 121, 364, 182, 91, 274, 137, 412, 206, 103, 310, 155, 466, 233, 700, 350, 175, 526, 263, 790, 395, 1186, 593, 1780, 890, 445, 1336, 668, 334, 167, 502, 251, 754, 377, 1132, 566, 283, 850, 425, 1276, 638, 319, 958, 479, 1438, 719, 2158, 1079, 3238, 1619, 4858, 2429, 7288, 3644, 1822, 911, 2734, 1367, 4102, 2051, 6154, 3077, 9232, 4616, 2308, 1154, 577, 1732, 866, 433, 1300, 650, 325, 976, 488, 244, 122, 61, 184, 92, 46, 23, 70, 35, 106, 53, 160, 80, 40, 20, 10, 5, 16, 8, **4**, **2**, **1** (, 4)]

48 | [48, 24, 12, 6, 3, 10, 5, 16, 8, **4**, **2**, **1** (, 4)]

49 | [49, 148, 74, 37, 112, 56, 28, 14, 7, 22, 11, 34, 17, 52, 26, 13, 40, 20, 10, 5, 16, 8, **4**, **2**, **1** (, 4)]

50 | [50, 25, 76, 38, 19, 58, 29, 88, 44, 22, 11, 34, 17, 52, 26, 13, 40, 20, 10, 5, 16, 8, **4**, **2**, **1** (, 4)]

51 | [51, 154, 77, 232, 116, 58, 29, 88, 44, 22, 11, 34, 17, 52, 26, 13, 40, 20, 10, 5, 16, 8, **4**, **2**, **1** (, 4)]

52 | [52, 26, 13, 40, 20, 10, 5, 16, 8, **4**, **2**, **1** (, 4)]

53 | [53, 160, 80, 40, 20, 10, 5, 16, 8, **4**, **2**, **1** (, 4)]

54 | [54, 27, 82, 41, 124, 62, 31, 94, 47, 142, 71, 214, 107, 322, 161, 484, 242, 121, 364, 182, 91, 274, 137, 412, 206, 103, 310, 155, 466, 233, 700, 350, 175, 526, 263, 790, 395, 1186, 593, 1780, 890, 445, 1336, 668, 334, 167, 502, 251, 754, 377, 1132, 566, 283, 850, 425, 1276, 638, 319, 958, 479, 1438, 719, 2158, 1079, 3238, 1619, 4858, 2429, 7288, 3644, 1822, 911, 2734, 1367, 4102, 2051, 6154, 3077, 9232, 4616, 2308, 1154, 577, 1732, 866, 433, 1300, 650, 325, 976, 488, 244, 122, 61, 184, 92, 46, 23, 70, 35, 106, 53, 160, 80, 40, 20, 10, 5, 16, 8, **4**, **2**, **1** (, 4)]

55 | [55, 166, 83, 250, 125, 376, 188, 94, 47, 142, 71, 214, 107, 322, 161, 484, 242, 121, 364, 182, 91, 274, 137, 412, 206, 103, 310, 155, 466, 233, 700, 350, 175, 526, 263, 790, 395, 1186, 593, 1780, 890, 445, 1336, 668, 334, 167, 502, 251, 754, 377, 1132, 566, 283, 850, 425, 1276, 638, 319, 958, 479, 1438, 719, 2158, 1079, 3238, 1619, 4858, 2429, 7288, 3644, 1822, 911, 2734, 1367, 4102, 2051, 6154, 3077, 9232, 4616, 2308, 1154, 577, 1732, 866, 433, 1300, 650, 325, 976, 488, 244, 122, 61, 184, 92, 46, 23, 70, 35, 106, 53, 160, 80, 40, 20, 10, 5, 16, 8, **4**, **2**, **1** (, 4)]

| | |
|---|---|
| 56 | [56, 28, 14, 7, 22, 11, 34, 17, 52, 26, 13, 40, 20, 10, 5, 16, 8, **4**, **2**, **1** (, 4)] |
| 57 | [57, 172, 86, 43, 130, 65, 196, 98, 49, 148, 74, 37, 112, 56, 28, 14, 7, 22, 11, 34, 17, 52, 26, 13, 40, 20, 10, 5, 16, 8, **4**, **2**, **1** (, 4)] |
| 58 | [58, 29, 88, 44, 22, 11, 34, 17, 52, 26, 13, 40, 20, 10, 5, 16, 8, **4**, **2**, **1** (, 4)] |
| 59 | [59, 178, 89, 268, 134, 67, 202, 101, 304, 152, 76, 38, 19, 58, 29, 88, 44, 22, 11, 34, 17, 52, 26, 13, 40, 20, 10, 5, 16, 8, **4**, **2**, **1** (, 4)] |
| 60 | [60, 30, 15, 46, 23, 70, 35, 106, 53, 160, 80, 40, 20, 10, 5, 16, 8, **4**, **2**, **1** (, 4)] |
| 61 | [61, 184, 92, 46, 23, 70, 35, 106, 53, 160, 80, 40, 20, 10, 5, 16, 8, **4**, **2**, **1** (, 4)] |
| 62 | [62, 31, 94, 47, 142, 71, 214, 107, 322, 161, 484, 242, 121, 364, 182, 91, 274, 137, 412, 206, 103, 310, 155, 466, 233, 700, 350, 175, 526, 263, 790, 395, 1186, 593, 1780, 890, 445, 1336, 668, 334, 167, 502, 251, 754, 377, 1132, 566, 283, 850, 425, 1276, 638, 319, 958, 479, 1438, 719, 2158, 1079, 3238, 1619, 4858, 2429, 7288, 3644, 1822, 911, 2734, 1367, 4102, 2051, 6154, 3077, 9232, 4616, 2308, 1154, 577, 1732, 866, 433, 1300, 650, 325, 976, 488, 244, 122, 61, 184, 92, 46, 23, 70, 35, 106, 53, 160, 80, 40, 20, 10, 5, 16, 8, **4**, **2**, **1** (, 4)] |
| 63 | [63, 190, 95, 286, 143, 430, 215, 646, 323, 970, 485, 1456, 728, 364, 182, 91, 274, 137, 412, 206, 103, 310, 155, 466, 233, 700, 350, 175, 526, 263, 790, 395, 1186, 593, 1780, 890, 445, 1336, 668, 334, 167, 502, 251, 754, 377, 1132, 566, 283, 850, 425, 1276, 638, 319, 958, 479, 1438, 719, 2158, 1079, 3238, 1619, 4858, 2429, 7288, 3644, 1822, 911, 2734, 1367, 4102, 2051, 6154, 3077, 9232, 4616, 2308, 1154, 577, 1732, 866, 433, 1300, 650, 325, 976, 488, 244, 122, 61, 184, 92, 46, 23, 70, 35, 106, 53, 160, 80, 40, 20, 10, 5, 16, 8, **4**, **2**, **1** (, 4)] |
| 64 | [64, 32, 16, 8, **4**, **2**, **1** (, 4)] |
| 65 | [65, 196, 98, 49, 148, 74, 37, 112, 56, 28, 14, 7, 22, 11, 34, 17, 52, 26, 13, 40, 20, 10, 5, 16, 8, **4**, **2**, **1** (, 4)] |
| 66 | [66, 33, 100, 50, 25, 76, 38, 19, 58, 29, 88, 44, 22, 11, 34, 17, 52, 26, 13, 40, 20, 10, 5, 16, 8, **4**, **2**, **1** (, 4)] |
| 67 | [67, 202, 101, 304, 152, 76, 38, 19, 58, 29, 88, 44, 22, 11, 34, 17, 52, 26, 13, 40, 20, 10, 5, 16, 8, **4**, **2**, **1** (, 4)] |
| 68 | [68, 34, 17, 52, 26, 13, 40, 20, 10, 5, 16, 8, **4**, **2**, **1** (, 4)] |
| 69 | [69, 208, 104, 52, 26, 13, 40, 20, 10, 5, 16, 8, **4**, **2**, **1** (, 4)] |
| 70 | [70, 35, 106, 53, 160, 80, 40, 20, 10, 5, 16, 8, **4**, **2**, **1** (, 4)] |

71 | [71, 214, 107, 322, 161, 484, 242, 121, 364, 182, 91, 274, 137, 412, 206, 103, 310, 155, 466, 233, 700, 350, 175, 526, 263, 790, 395, 1186, 593, 1780, 890, 445, 1336, 668, 334, 167, 502, 251, 754, 377, 1132, 566, 283, 850, 425, 1276, 638, 319, 958, 479, 1438, 719, 2158, 1079, 3238, 1619, 4858, 2429, 7288, 3644, 1822, 911, 2734, 1367, 4102, 2051, 6154, 3077, 9232, 4616, 2308, 1154, 577, 1732, 866, 433, 1300, 650, 325, 976, 488, 244, 122, 61, 184, 92, 46, 23, 70, 35, 106, 53, 160, 80, 40, 20, 10, 5, 16, 8, **4, 2, 1** (, 4)]

72 | [72, 36, 18, 9, 28, 14, 7, 22, 11, 34, 17, 52, 26, 13, 40, 20, 10, 5, 16, 8, **4, 2, 1** (, 4)]

73 | [73, 220, 110, 55, 166, 83, 250, 125, 376, 188, 94, 47, 142, 71, 214, 107, 322, 161, 484, 242, 121, 364, 182, 91, 274, 137, 412, 206, 103, 310, 155, 466, 233, 700, 350, 175, 526, 263, 790, 395, 1186, 593, 1780, 890, 445, 1336, 668, 334, 167, 502, 251, 754, 377, 1132, 566, 283, 850, 425, 1276, 638, 319, 958, 479, 1438, 719, 2158, 1079, 3238, 1619, 4858, 2429, 7288, 3644, 1822, 911, 2734, 1367, 4102, 2051, 6154, 3077, 9232, 4616, 2308, 1154, 577, 1732, 866, 433, 1300, 650, 325, 976, 488, 244, 122, 61, 184, 92, 46, 23, 70, 35, 106, 53, 160, 80, 40, 20, 10, 5, 16, 8, **4, 2, 1** (, 4)]

74 | [74, 37, 112, 56, 28, 14, 7, 22, 11, 34, 17, 52, 26, 13, 40, 20, 10, 5, 16, 8, **4, 2, 1** (, 4)]

75 | [75, 226, 113, 340, 170, 85, 256, 128, 64, 32, 16, 8, **4, 2, 1** (, 4)]

76 | [76, 38, 19, 58, 29, 88, 44, 22, 11, 34, 17, 52, 26, 13, 40, 20, 10, 5, 16, 8, **4, 2, 1** (, 4)]

77 | [77, 232, 116, 58, 29, 88, 44, 22, 11, 34, 17, 52, 26, 13, 40, 20, 10, 5, 16, 8, **4, 2, 1** (, 4)]

78 | [78, 39, 118, 59, 178, 89, 268, 134, 67, 202, 101, 304, 152, 76, 38, 19, 58, 29, 88, 44, 22, 11, 34, 17, 52, 26, 13, 40, 20, 10, 5, 16, 8, **4, 2, 1** (, 4)]

79 | [79, 238, 119, 358, 179, 538, 269, 808, 404, 202, 101, 304, 152, 76, 38, 19, 58, 29, 88, 44, 22, 11, 34, 17, 52, 26, 13, 40, 20, 10, 5, 16, 8, **4, 2, 1** (, 4)]

80 | [80, 40, 20, 10, 5, 16, 8, **4, 2, 1** (, 4)]

81 | [81, 244, 122, 61, 184, 92, 46, 23, 70, 35, 106, 53, 160, 80, 40, 20, 10, 5, 16, 8, **4, 2, 1** (, 4)]

82 | [82, 41, 124, 62, 31, 94, 47, 142, 71, 214, 107, 322, 161, 484, 242, 121, 364, 182, 91, 274, 137, 412, 206, 103, 310, 155, 466, 233, 700, 350, 175, 526, 263, 790, 395, 1186, 593, 1780, 890, 445, 1336, 668, 334, 167, 502, 251, 754, 377, 1132, 566, 283, 850, 425, 1276, 638, 319, 958, 479, 1438, 719, 2158, 1079, 3238, 1619, 4858, 2429, 7288, 3644, 1822, 911, 2734, 1367, 4102, 2051, 6154, 3077, 9232, 4616, 2308, 1154, 577, 1732, 866, 433, 1300, 650, 325, 976, 488, 244, 122, 61, 184, 92, 46, 23, 70, 35, 106, 53, 160, 80, 40, 20, 10, 5, 16, 8, **4, 2, 1** (, 4)]

83 | [83, 250, 125, 376, 188, 94, 47, 142, 71, 214, 107, 322, 161, 484, 242, 121, 364, 182, 91, 274, 137, 412, 206, 103, 310, 155, 466, 233, 700, 350, 175, 526, 263, 790, 395, 1186, 593, 1780, 890, 445, 1336, 668, 334, 167, 502, 251, 754, 377, 1132, 566, 283, 850, 425, 1276, 638, 319, 958, 479, 1438, 719, 2158, 1079, 3238, 1619, 4858, 2429, 7288, 3644, 1822, 911, 2734, 1367, 4102, 2051, 6154, 3077, 9232, 4616, 2308, 1154, 577, 1732, 866, 433, 1300, 650, 325, 976, 488, 244, 122, 61, 184, 92, 46, 23, 70, 35, 106, 53, 160, 80, 40, 20, 10, 5, 16, 8, **4, 2, 1** (, 4)]

84 | [84, 42, 21, 64, 32, 16, 8, **4, 2, 1** (, 4)]

85 | [85, 256, 128, 64, 32, 16, 8, **4, 2, 1** (, 4)]

86 | [86, 43, 130, 65, 196, 98, 49, 148, 74, 37, 112, 56, 28, 14, 7, 22, 11, 34, 17, 52, 26, 13, 40, 20, 10, 5, 16, 8, **4, 2, 1** (, 4)]

87 | [87, 262, 131, 394, 197, 592, 296, 148, 74, 37, 112, 56, 28, 14, 7, 22, 11, 34, 17, 52, 26, 13, 40, 20, 10, 5, 16, 8, **4, 2, 1** (, 4)]

88 | [88, 44, 22, 11, 34, 17, 52, 26, 13, 40, 20, 10, 5, 16, 8, **4, 2, 1** (, 4)]

89 | [89, 268, 134, 67, 202, 101, 304, 152, 76, 38, 19, 58, 29, 88, 44, 22, 11, 34, 17, 52, 26, 13, 40, 20, 10, 5, 16, 8, **4, 2, 1** (, 4)]

90 | [90, 45, 136, 68, 34, 17, 52, 26, 13, 40, 20, 10, 5, 16, 8, **4, 2, 1** (, 4)]

91 | [91, 274, 137, 412, 206, 103, 310, 155, 466, 233, 700, 350, 175, 526, 263, 790, 395, 1186, 593, 1780, 890, 445, 1336, 668, 334, 167, 502, 251, 754, 377, 1132, 566, 283, 850, 425, 1276, 638, 319, 958, 479, 1438, 719, 2158, 1079, 3238, 1619, 4858, 2429, 7288, 3644, 1822, 911, 2734, 1367, 4102, 2051, 6154, 3077, 9232, 4616, 2308, 1154, 577, 1732, 866, 433, 1300, 650, 325, 976, 488, 244, 122, 61, 184, 92, 46, 23, 70, 35, 106, 53, 160, 80, 40, 20, 10, 5, 16, 8, **4, 2, 1** (, 4)]

92 | [92, 46, 23, 70, 35, 106, 53, 160, 80, 40, 20, 10, 5, 16, 8, **4, 2, 1** (, 4)]

93 | [93, 280, 140, 70, 35, 106, 53, 160, 80, 40, 20, 10, 5, 16, 8, **4, 2, 1** (, 4)]

94 | [94, 47, 142, 71, 214, 107, 322, 161, 484, 242, 121, 364, 182, 91, 274, 137, 412, 206, 103, 310, 155, 466, 233, 700, 350, 175, 526, 263, 790, 395, 1186, 593, 1780, 890, 445, 1336, 668, 334, 167, 502, 251, 754, 377, 1132, 566, 283, 850, 425, 1276, 638, 319, 958, 479, 1438, 719, 2158, 1079, 3238, 1619, 4858, 2429, 7288, 3644, 1822, 911, 2734, 1367, 4102, 2051, 6154, 3077, 9232, 4616, 2308, 1154, 577, 1732, 866, 433, 1300, 650, 325, 976, 488, 244, 122, 61, 184, 92, 46, 23, 70, 35, 106, 53, 160, 80, 40, 20, 10, 5, 16, 8, **4, 2, 1** (, 4)]

| | |
|---|---|
| 95 | [95, 286, 143, 430, 215, 646, 323, 970, 485, 1456, 728, 364, 182, 91, 274, 137, 412, 206, 103, 310, 155, 466, 233, 700, 350, 175, 526, 263, 790, 395, 1186, 593, 1780, 890, 445, 1336, 668, 334, 167, 502, 251, 754, 377, 1132, 566, 283, 850, 425, 1276, 638, 319, 958, 479, 1438, 719, 2158, 1079, 3238, 1619, 4858, 2429, 7288, 3644, 1822, 911, 2734, 1367, 4102, 2051, 6154, 3077, 9232, 4616, 2308, 1154, 577, 1732, 866, 433, 1300, 650, 325, 976, 488, 244, 122, 61, 184, 92, 46, 23, 70, 35, 106, 53, 160, 80, 40, 20, 10, 5, 16, 8, **4, 2, 1** (, 4)] |
| 96 | [96, 48, 24, 12, 6, 3, 10, 5, 16, 8, **4, 2, 1** (, 4)] |
| 97 | [97, 292, 146, 73, 220, 110, 55, 166, 83, 250, 125, 376, 188, 94, 47, 142, 71, 214, 107, 322, 161, 484, 242, 121, 364, 182, 91, 274, 137, 412, 206, 103, 310, 155, 466, 233, 700, 350, 175, 526, 263, 790, 395, 1186, 593, 1780, 890, 445, 1336, 668, 334, 167, 502, 251, 754, 377, 1132, 566, 283, 850, 425, 1276, 638, 319, 958, 479, 1438, 719, 2158, 1079, 3238, 1619, 4858, 2429, 7288, 3644, 1822, 911, 2734, 1367, 4102, 2051, 6154, 3077, 9232, 4616, 2308, 1154, 577, 1732, 866, 433, 1300, 650, 325, 976, 488, 244, 122, 61, 184, 92, 46, 23, 70, 35, 106, 53, 160, 80, 40, 20, 10, 5, 16, 8, **4, 2, 1** (, 4)] |
| 98 | [98, 49, 148, 74, 37, 112, 56, 28, 14, 7, 22, 11, 34, 17, 52, 26, 13, 40, 20, 10, 5, 16, 8, **4, 2, 1** (, 4)] |
| 99 | [99, 298, 149, 448, 224, 112, 56, 28, 14, 7, 22, 11, 34, 17, 52, 26, 13, 40, 20, 10, 5, 16, 8, **4, 2, 1** (, 4)] |
| 100 | [100, 50, 25, 76, 38, 19, 58, 29, 88, 44, 22, 11, 34, 17, 52, 26, 13, 40, 20, 10, 5, 16, 8, **4, 2, 1** (, 4)] |

## Some Tough Problems Made Simple

Here again, we are faced with a problem that asks us to work with inordinately large numbers.

*Find the units digit for the following sum:* $13^{25} + 4^{81} + 5^{411}$.

Again, when members of your audience are faced with this problem, they will probably attempt to solve this problem by using their calculators. This is a formidable task, and an error can often be expected! The beauty in this problem is not simply getting the answer, rather finding the *path* to the answer. Let's utilize the strategy that requires us to look for a pattern. We must examine the patterns that exist in the powers of three different sets of numbers. Practice in doing this will help familiarize you with the cyclical pattern for the final digits of the powers of numbers.

For powers of 13, we obtain

$$13^1 = 1\underline{3} \qquad 13^5 = 371,29\underline{3}$$
$$13^2 = 16\underline{9} \qquad 13^6 = 4,826,80\underline{9}$$
$$13^3 = 2,19\underline{7} \qquad 13^7 = 62,748,51\underline{7}$$
$$13^4 = 28,56\underline{1} \qquad 13^8 = 815,730,72\underline{1}$$

The units digits for powers of 13 repeat as 3, 9, 7, 1, 3, 9, 7, 1, ... in cycles of 4. Thus $13^{25}$ has the same units digit as $13^1$ or 3.

For powers of 4, we obtain

$$4^1 = \underline{4} \qquad 4^5 = 1,02\underline{4}$$
$$4^2 = 1\underline{6} \qquad 4^6 = 4,09\underline{6}$$
$$4^3 = 6\underline{4} \qquad 4^7 = 16,38\underline{4}$$
$$4^4 = 25\underline{6} \qquad 4^8 = 65,53\underline{6}$$

The units digits for powers of 4 repeat as 4, 6, 4, 6, 4, 6, ... in cycles of 2. Thus $4^{81}$ has the same units digit as $4^1$ which is 4.

The units digit for powers of 5 must be the numeral 5 (i.e., $\underline{5}$, $2\underline{5}$, $12\underline{5}$, $62\underline{5}$, etc.).

The sum we are looking for is $3 + 4 + 5 = 12$, which has a units digit of 2.

In case anybody actually want to see what the value of this addition is, we offer it here.

$(13^{25} + 4^{81} + 5^{411} = 18909140209225186878994290201593514880713960898675736647889467487033282949695732250306065597055735336465124719275168298532084162104454835525011318606705812949230644849953763685246250187369017353959030115461205773438385108215717621322345025635355801849375382828439521674804517901168473972\underline{2})$.

## An Audience Challenge

Here is another problem that will lead us to a clever "aha" solution, which could enthrall your audience.

*What is the quotient of 1 divided by 500,000,000,000?*

This can be restated as, find the value of $\frac{1}{500,000,000,000}$?

This problem cannot be done on the calculator, since the answer will contain more places than the display would permit. It can be done manually, although the computation often leads to an error due to the large number of 0's in the answer. We might, however, examine the answers we obtain by starting with a small divisor, increasing the divisor, and seeing if a usable pattern emerges. (See Figure 1.18.)

| | Number of 0's after the 5 | Quotient | Number of 0's after the decimal and before the 2 |
|---|---|---|---|
| $1 \div 5$ | 0 | .2 | 0 |
| $1 \div 50$ | 1 | .02 | 1 |
| $1 \div 500$ | 2 | .002 | 2 |
| $1 \div 5000$ | 3 | .0002 | 3 |
| . . . | | | |
| $1 \div 500000000000$ | 11 | .000000000002 | 11 |

**Figure 1.18**

The correct answer is now easily found. The number of 0's after the decimal point and before the 2 is the same as the number of zeros in the divisor, $\frac{1}{500,000,000,000} = 2 \cdot 10^{-12} = 0.2 \cdot 10^{-11}$.

## Impressing the Audience with a Surprise Pattern

Begin by posing the following problem, which on the surface seems rather harmless but could get a bit cumbersome.

*What is the sum of $1^3 + 2^3 + 3^3 + 4^3 + \cdots + 9^3 + 10^3$?*

One could actually find the cube of all the integers from 1 to 10, and then take the sum. If carefully done (with the aid of a calculator), this should yield the correct answer. However, if we do not have a calculator handy, the multiplication and addition could prove quite cumbersome and messy! Let's see how we might solve the problem by searching for a pattern. Let's organize our data with a table:

| | | | |
|---|---|---|---|
| $1^3$ | $= (1)$ | $= 1$ | $= 1^2$ |
| $1^3 + 2^3$ | $= (1 + 8)$ | $= 9$ | $= 3^2$ |
| $1^3 + 2^3 + 3^3$ | $= (1 + 8 + 27)$ | $= 36$ | $= 6^2$ |
| $1^3 + 2^3 + 3^3 + 4^3$ | $= (1 + 8 + 27 + 64)$ | $= 100$ | $= 10^2$ |

Notice that the number bases in the final column (namely 1, 3, 6, 10, ...) are our now-familiar *triangular numbers*. The $n$th triangular number is formed by taking the sum of the first $n$ integers. That is, the first triangular number is 1; the second triangular number is 3 = (1 + 2); the third triangular number is 6 = (1 + 2 + 3); the fourth triangular number is 10 = (1 + 2 + 3 + 4), and so on.

Thus, we can rewrite our problem as follows:

| | | |
|---|---|---|
| $1^3$ | $= (1)^2$ | $= 1^2 = 1$ |
| $1^3 + 2^3$ | $= (1 + 2)^2$ | $= 32 = 9$ |
| $1^3 + 2^3 + 3^3$ | $= (1 + 2 + 3)2$ | $= 62 = 36$ |
| $1^3 + 2^3 + 3^3 + \cdots + 9^3 + 10^3$ | $= (1 + 2 + 3 + \cdots + 9 + 10)^2$ | $= 55^2 = 3025$ |

By this point your audience should have gotten a "feel" for the advantage of looking for a pattern in solving a problem. It may take some effort to find a pattern, but when one is discovered, it not only simplifies the problem greatly, but also once again demonstrates the beauty of mathematics.

## A Tough Challenge

It is not uncommon that some people like to be challenged with an arithmetic problem, so here we present this kind of entertainment.

Consider the two products: $158 \times 23$ and $79 \times 46$. In each case the product is 3,634. The challenge is to rearrange the digits keeping a first product having a three-digit and two-digit number and a second product having a pair of two-digit numbers, so that the largest possible product will be the same for each group.

Typically, most people will approach this in a "trial and error" fashion and with some luck stumble upon a correct answer. However, this will not satisfy most problem solvers. A clever person might try to reverse the digits of each of the two second-numbers of the given products and find that they have arrived at two equal products, thinking that these may be the largest possible equal products. Delighted to have achieved equal products as follows: $158 \times 32 = 5056$, and $79 \times 64 = 5056$, they might want to conclude that the original challenge has been achieved.

Unfortunately, this product is not the largest possible. A much larger product 5568 can be obtained by multiplying $174 \times 32$, and by multiplying $96 \times 58$. A solution that will require a fair amount of effort.

## Thou Shalt Not Divide by Zero

Every mathematician knows that division by zero is forbidden. As a matter of fact, on the list of commandments in mathematics this must certainly be at the top. However, we can refer to this as the "eleventh commandment." But why is division by zero not permissible? We in mathematics pride ourselves with the order and beauty in which everything in the realm of mathematics falls neatly into place. When something arises that could spoil that order, we simple *define* it to suit our needs. This is precisely what happens with division by zero. One gets a much greater insight into the nature of mathematics by explaining why these "rules" are set forth. So, let's give this "commandment" some meaning. For many in your audience this could be a long-awaited revelation.

Consider the quotient $\frac{n}{0}$. Without acknowledging the division-by-zero commandment, let us speculate (i.e., guess) what the quotient might be. Let us say it is $p$. In that case we could check by multiplying $0 \cdot p$ to see if it equals $n$ as would have to be the case for the division to be correct. We know that $0 \cdot p \neq a$, since $0 \cdot p = 0$. So there is no number, $p$, that can take on the quotient to this division. For that reason, we define division by zero to be invalid.

A more convincing case for defining away division by zero is to show how it can lead to a contradiction of an accepted fact, namely, that $1 \neq 2$. We will show that when division by zero is acceptable, then $1 = 2$; clearly an absurdity!

Here is the "proof" that $1 = 2$:

$$\text{Let } a = b$$
$$\text{Then } a^2 = ab \qquad \text{[multiplying both sides by } a\text{]}$$
$$a^2 - b^2 = ab - b^2 \qquad \text{[subtracting } b^2 \text{ from both sides]}$$
$$(a-b)(a+b) = b(a-b) \qquad \text{[factoring]}$$
$$a + b = b \qquad \text{[dividing by } (a-b)\text{]}$$
$$2b = b \qquad \text{[replace } a \text{ by } b\text{ ]}$$
$$2 = 1 \qquad \text{[divide both sides by } b\text{]}$$

In the step where we divided by $(a - b)$, we actually divided by zero, because $a = b$, so that $a - b = 0$. That ultimately led us to an absurd result, leaving us with no option other than to prohibit division by zero. By taking the time to witness this rule about division by zero, your audience will have a much better appreciation for the nature of mathematics. Then, of course, they will be able to enjoy it, because they are gently being shown not to simply accept everything as "in mathematics it must be just so." A critical eye onto the subject matter is healthy!

Sometimes, a situation that might involve the 11th commandment is well camouflaged. Take, for example, the equation $5x + 12 = 6x + 30$. This can be rewritten as $5x - 30 = 6x - 12$, which by factoring both sides gives us $5(3x - 6) = 2(3x - 6)$. When we divide both sides of the equation by $3x - 6$, we get the absurd result that $5 = 2$. The question then is, where has a mistake been made in our algebraic solution? You will relieve your audience by explaining to them that had we solved the equation in the original format, we would have found that $x = 2$. In that case, the expression $3x - 6$ would equal $6 - 6 = 0$. Lo and behold, we once again divided by zero and violated the 11th commandment.

## What's Wrong Here?

Just to keep your audience consistently sharp, we offer them another conundrum.

In elementary algebra we learn that $\frac{a^3-b^3}{a-b} = \frac{(a^2+ab+b^2)(a-b)}{a-b} = a^2+ab+b^2$. When $a = b$, we have a problem in that the equality does not hold. On the left-hand side, we get $\frac{0}{0} = 0$, and on the right-hand side, we get $1 + 1 + 1 = 3$. This would imply that $0 = 3$, which is ridiculous. Here is another example to fortify our previously determined "11th commandment!" By now, the audience ought to be thoroughly comfortable with why division by zero is not acceptable in mathematics.

## An Infinite Series Fallacy

Here is one that will leave many members of your audience somewhat baffled. Yet the "answer" is a bit subtle and may be require some more mature thought. By ignoring the notion of a convergent series[8] we get the following dilemma:

$$\begin{aligned} \text{Let } S &= 1-1+1-1+1-1+1-1+\cdots \\ &= (1-1)+(1-1)+(1-1)+(1-1)+\cdots \\ &= 0+0+0+0+\cdots \\ &= 0 \end{aligned}$$

However, if we were to group this differently, we would get:

$$\begin{aligned} \text{Let } S &= 1-1 \ +1-1+1-1+1-1+\cdots \\ &= 1-(1-1)-(1-1)-(1-1)-\cdots \\ &= 1-0-0-0-\cdots \\ &= 1 \end{aligned}$$

Therefore, since in the first case, $S = 1$, and in the second case, $S = 0$, we could conclude that $1 = 0$. What's wrong with this argument? If this hasn't upset you enough, consider the following argument: Let

$$S = 1 + 2 + 4 + 8 + 16 + 32 + 64 + \cdots \tag{1}$$

Here $S$ is clearly positive.

Also,

$$S - 1 = 2 + 4 + 8 + 16 + 32 + 64 + \cdots \tag{2}$$

Now by multiplying both sides of equation (1) by 2, we obtain

$$2S = 2 + 4 + 8 + 16 + 32 + 64 + \cdots \tag{3}$$

Substituting equation (2) into equation (3) gives us

$$2S = S - 1$$

From which we can conclude that $S = -1$.

This would have us to conclude that $-1$ is positive, since we established earlier that $S$ was positive.

To clarify the last fallacy, you might want to compare the following correct form of a convergent series: Let $S = 1 + \frac{1}{2} + \frac{1}{4} + \frac{1}{8} + \frac{1}{16}$. We then have $2S = 2 + 1 + \frac{1}{2} + \frac{1}{4} + \frac{1}{8} + \frac{1}{16}$.

Then $2S = 2 + S$, and $S = 2$, which is true. The difference lies in the notion of a convergent series as is this last one, while the earlier ones were not convergent, and therefore, do not allow for the assumptions we made.

## A Baffling Solution to an Arithmetic Calculation

Typically, an arithmetic sequence can be easily summed. There are, however, sequences that are a little strange in their presentation and then as a result are overwhelming as one tries to work with them. Yet they can be entertaining especially after the solution has been presented. One example of this situation is with the series:

$$5 \times 7 \times 9 + 7 \times 9 \times 11 + 9 \times 11 \times 13 + 11 \times 13 \times 15 + 13 \times 15 \times 17 + \cdots$$

One challenge would be to find the ninth term of this series. Then, after finding the ninth term, determining the sum of the entire series up to the ninth term. Following the pattern that should be pretty obvious at this point, determining the ninth term should be rather simple, namely, $21 \times 23 \times 25$. However, obtaining the sum of the series is a bit more complicated. There are various ways in which this can be done; however, we will use a rather strange device for entertainment purposes. We will take the last term and multiply it by the first factor of the next term, leaving us with $21 \times 23 \times 25 \times 27$. We then take the first term and multiply it by the last factor of the previous term (if there were one) to get $3 \times 5 \times 7 \times 9$. We then subtract these two products, so that we have: $21 \times 23 \times 25 \times 27 - 3 \times 5 \times 7 \times 9 = 326{,}025 - 945 = 325{,}080$. We will then divide this number by $2 \times (3 + 1) = 8$, which was obtained by the following rule: the 2 is the difference between factors, and 3 as a number of factors. Therefore, the series some is $325{,}080 \div 8 = 40{,}635$. As confusing and unusual as this solution may be, a motivated receiver of this procedure might want to determine the justification. There is then another form of entertainment!

## An Amazing Phenomenon

As we close this chapter, we present a truly amazing numerical relationship. This one will clearly baffle your audience leaving them with a true wonder about the field of arithmetic. We begin with the equation of numbers equal to 0.

$$123789^2 + 561945^2 + 642864^2 - 242868^2 - 761943^2 - 323787^2 = 0$$

Looking at this relationship there is nothing particularly strange other than the fact that the numbers are rather large. However, when we delete the 100-thousands place (the left-most digit) from each number, we get the following relationship which remains equal to 0:

$$23789^2 + 61945^2 + 42864^2 - 42868^2 - 61943^2 - 23787^2 = 0$$

When we repeat this process by deleting the left-most digit of each number, we are left with another relationship which is again equal to 0:

$$3789^2 + 1945^2 + 2864^2 - 2868^2 - 1943^2 - 3787^2 = 0$$

When we continue this process of deleting the leftmost digit in each case the result remains 0:

$$789^2 + 945^2 + 864^2 - 868^2 - 943^2 - 787^2 = 0$$
$$89^2 + 45^2 + 64^2 - 68^2 - 43^2 - 87^2 = 0$$
$$9^2 + 5^2 + 4^2 - 8^2 - 3^2 - 7^2 = 0$$

At this point, some people may think that the problem was rigged, as well it might have been. However, we can take the same sequence and repeat the process. But this time deleting the units digit sequentially from each of the numbers, and once again noticing that the equation is equal to 0.

$$123789^2 + 561945^2 + 642864^2 - 242868^2 - 761943^2 - 323787^2 = 0$$
$$12378^2 + 56194^2 + 64286^2 - 24286^2 - 76194^2 - 32378^2 = 0$$
$$1237^2 + 5619^2 + 6428^2 - 2428^2 - 7619^2 - 3237^2 = 0$$
$$123^2 + 561^2 + 642^2 - 242^2 - 761^2 - 323^2 = 0$$
$$12^2 + 56^2 + 64^2 - 24^2 - 76^2 - 32^2 = 0$$
$$1^2 + 5^2 + 6^2 - 2^2 - 7^2 - 3^2 = 0$$

If your audience is not yet impressed enough, then you have an opportunity to show off a bit more by combining the two types of deletions that we have done above, simultaneously! That is, that we will remove the rightmost and leftmost digits at the same time and once again amaze the audience while achieving a sum of zero with each pair of deletions.

$$123789^2 + 561945^2 + 642864^2 - 242868^2 - 761943^2 - 323787^2 = 0$$
$$2378^2 + 6194^2 + 4286^2 - 4286^2 - 6194^2 - 2378^2 = 0$$

$$37^2 + 19^2 + 28^2 - 28^2 - 19^2 - 37^2 = 0$$

With this wild challenge we figured to have occupied your audience sufficiently and can move on beyond arithmetic to the other areas of mathematics, which should also provide a vast variety of entertainments.

# Endnotes

[1] We will discuss Kaprekar numbers on page 80.

[2] Proper divisors are all the divisors, or factors, of the number except the number itself. For example, the proper divisors of 6 are 1, 2, and 3, but not 6.

[3] For a proof that this relationship holds as started, see Ross Honsberger, *Ingenuity in Mathematics*, New York: Random House, 1970, pp. 147–156.

[4] E. A. Maxwell, *Fallacies in Mathematics*, Cambridge: University Press, 1963.

[5] Kaprekar announced it at the Madras Mathematical Conference in 1949. He published the result in the paper "*Problems involving reversal of digits*" in *Scripta Mathematica* in 1953; see also Kaprekar, D. R.: An Interesting Property of the Number 6174. *Scripta Math.* 15(1955), 244–245.

[6] If you choose a six-digit number (not one with all of the same digits), then there are also three Kaprekar constants: 549945, 631764, and 420876.
　　Two of length 1: [549945 (,549945)], [631764 (,631764)]
　　and one of length 7: [420876, 851742, 750843, 840852, 860832, 862632, 642654 (, 420876)].
　　If you choose a seven-digit number (not one with all of the same digits), then there is only one Kaprekar constant: 7509843. There is a loop of length 8: [7509843, 9529641, 8719722, 8649432, 7519743, 8429652, 7619733, 8439552 (, 7509843)].

[7] In a subtraction, the number in the *subtrahend* is subtracted from the number in the *minuend* to get the result, referred to as the *difference*.

[8] In simple terms, a series converges if it appears to be approaching a specific finite sum. For example, the series $1 + \frac{1}{2} + \frac{1}{4} + \frac{1}{8} + \frac{1}{16} + \frac{1}{32} + \cdots$ converges to 2, while the series $1 + \frac{1}{2} + \frac{1}{3} + \frac{1}{4} + \frac{1}{5} + \frac{1}{6} + \cdots$ does not converge to any finite sum but continues to grow indefinitely.

# Chapter 2

# The Fun of Logical Reasoning

The focus of logical reasoning is an integral aspect of all branches of mathematics. Logical reasoning is often seen as a key element of genuine problem-solving skills. The topics selected to exhibit logical reasoning can not only vary greatly, but can also exhibit the delight that one can have with some of the amazing aspects of mathematics. The examples that we will encounter throughout this chapter will manifest themselves in such areas as probability, clever strategies for solving problems, arriving at unexpected phenomena, and other entertaining aspects which exhibit some logical reasoning. Logical reasoning also has limits which could be entertaining. On the one hand, logical reasoning is often key to solving problems. Yet, on the other hand, one needs to be careful and not abuse logical reasoning. For example, making generalizations based on what appear to be convincing patterns can also be misleading. To see such surprising and unexpected disappointments to proper expectations can also be a form of entertainment, and also a caution for future conjectures. Let's begin by considering the following example.

## Making Mistaken Generalizations

It is easy to try to make a generalization when a pattern appears to be consistent. However, patterns can be consistent up to a certain point, after which an inconsistency can disturb the pattern. This is something that is too often not mentioned in the school curriculum. Perhaps teachers don't want to upset students by showing them that there can be such unexpected inconsistencies in mathematics. Let's take a look at one such example by considering the following question.

*Can every odd number greater than 1 be expressed as the sum of a (power of 2) + (a prime number)?*

It is not unusual that one tries to see if the first several examples hold up to the question asked. In the following pattern, we notice that it appears to hold true for all the odd numbers from 1 to 125, but then at the number 127 it does not hold true. For most people, this is truly shocking but presented properly this can be quite entertaining as well as providing a signal of caution when making generalizations. To continue on from 129, it will continue to hold true for a while. Your audience may wish to see how far they can go before they reach another "stumbling block."

$$3 = 2^0 + 2$$
$$5 = 2^1 + 3$$
$$7 = 2^2 + 3$$
$$9 = 2^2 + 5$$
$$11 = 2^3 + 3$$
$$13 = 2^3 + 5$$
$$15 = 2^3 + 7$$
$$17 = 2^2 + 13$$
$$19 = 2^4 + 3$$
$$\vdots$$
$$51 = 2^5 + 19$$
$$\vdots$$

$$125 = 2^6 + 61$$
$$127 = ?$$
$$129 = 2^5 + 97$$
$$131 = 2^7 + 3$$

The pattern then continues on and doesn't falter until the number 149. This conjecture was originally proposed by the French mathematician Alphonse de Polignac (1817–1890) and continues to falter with the following numbers: 251, 331, 337, 373, and 509. From the time when this question arose, it has been proved that there are an infinite number of such "failures" of this conjecture, of which the number 2,999,999 is one such failure.

Just for amusement, it might be nice to give another example of something that looks like a pattern that could go on infinitely long, and yet surprisingly, can halt abruptly. Here is another illustration that appears to produce a delightful pattern, but cannot be extended beyond a certain point. Admire the equalities that we find below by taking certain numbers to powers of 1, 2, 3, 4, 5, 6, and 7,

$$1^0 + 13^0 + 28^0 + 70^0 + 82^0 + 124^0 + 139^0 + 151^0 = 4^0 + 7^0 + 34^0 + 61^0 + 91^0 + 118^0 + 145^0 + 148^0$$
$$1^1 + 13^1 + 28^1 + 70^1 + 82^1 + 124^1 + 139^1 + 151^1 = 4^1 + 7^1 + 34^1 + 61^1 + 91^1 + 118^1 + 145^1 + 148^1$$
$$1^2 + 13^2 + 28^2 + 70^2 + 82^2 + 124^2 + 139^2 + 151^2 = 4^2 + 7^2 + 34^2 + 61^2 + 91^2 + 118^2 + 145^2 + 148^2$$
$$1^3 + 13^3 + 28^3 + 70^3 + 82^3 + 124^3 + 139^3 + 151^3 = 4^3 + 7^3 + 34^3 + 61^3 + 91^3 + 118^3 + 145^3 + 148^3$$
$$1^4 + 13^4 + 28^4 + 70^4 + 82^4 + 124^4 + 139^4 + 151^4 = 4^4 + 7^4 + 34^4 + 61^4 + 91^4 + 118^4 + 145^4 + 148^4$$
$$1^5 + 13^5 + 28^5 + 70^5 + 82^5 + 124^5 + 139^5 + 151^5 = 4^5 + 7^5 + 34^5 + 61^5 + 91^5 + 118^5 + 145^5 + 148^5$$
$$1^6 + 13^6 + 28^6 + 70^6 + 82^6 + 124^6 + 139^6 + 151^6 = 4^6 + 7^6 + 34^6 + 61^6 + 91^6 + 118^6 + 145^6 + 148^6$$
$$1^7 + 13^7 + 28^7 + 70^7 + 82^7 + 124^7 + 139^7 + 151^7 = 4^7 + 7^7 + 34^7 + 61^7 + 91^7 + 118^7 + 145^7 + 148^7$$

From these 7 examples, one could easily form the following conclusion, namely that, for the natural number $n$, the following condition should hold:

$$1^n + 13^n + 28^n + 70^n + 82^n + 124^n + 139^n + 151^n$$
$$= 4^n + 7^n + 34^n + 61^n + 91^n + 118^n + 145^n + 148^n$$

To provide your audience with some more information about the above 8 examples, we offer the 7 values of $n$ and obtain the sums as shown in the following table:

| $n$ | Sums |
|---|---:|
| 0 | 8 |
| 1 | 608 |
| 2 | 70,076 |
| 3 | 8,953,712 |
| 4 | 1,199,473,412 |
| 5 | 165,113,501,168 |
| 6 | 23,123,818,467,476 |
| 7 | 3,276,429,220,606,352 |

To make a generalization from this pattern would be expected. However, at the same time, it would also be a marvelous mistake. This mistake does not manifest itself until we consider the next case, namely, where $n = 8$. We notice that the two sums that we get are no longer the same: $1^8 + 13^0 + 28^0 + 70^0 + 82^8 + 124^8 + 139^8 + 151^8 = 468,150,771,944,932,292$, while $4^8 + 7^8 + 34^8 + 61^8 + 91^8 + 118^8 + 145^8 + 148^8 = 468,087,218,970,647,492$.

As a matter of fact, the difference between these two sums is

$$468,150,771,944,932,292 - 468,087,218,970,647,492$$
$$= 63,552,974,284,800$$

As $n$ increases, so does the difference between the two sums. For $n = 20$, the difference is 3,388,331,687,715,737,094,794,416,650,060, 343,026,048,000.

Therefore, to avoid such mistakes, one must be sure to prove a generalization before accepting it inductively. Properly presented, these false generalizations can be entertaining in a way that shows that not everything that looks like it's predictable is in fact not that.

## Some Cautions

Now that we have seen that generalizations need to be scrutinized, let's turn our attention to the logical reasoning that often makes a seemingly very difficult problem trivial. We begin with a problem that might be counterintuitive, and which we hope will open your audience's scope to a broader vision.

Offer them the following challenge. Supposing you are offered either a gallon jug filled with half-dollar coins or the same size jug filled with dimes. Initially one would think that you would get 5 times as much money out of the jug holding the half-dollar coins. However, because the half-dollar coins are larger, the jug will hold fewer half-dollar coins than they would hold dimes. It is estimated that there would be 6 times as many dimes as half-dollar coins which would indicate that you have approximately 20% more money in the jug holding the dimes. This is another example of using logical reasoning and one that gives you a sense that counterintuitivity should not be ignored.

## Logic in Arithmetic

Here is an arithmetic question that requires a logical approach. If a store sells a variety of items all at the same price and each item costs more than $.20, how many items can you buy for $3.41?

Here is where your audience needs to see which numbers are factors of 341. If they recall from Chapter 1, the technique for testing for divisibility by 11, they would realize that the difference of the sums of the alternate digits of 341 is zero, which is divisible by 11, and therefore 11 is a factor of 341. Consequently, you could by 11 items at the cost of $.31 per item. Here, we see how a logical approach buttressed by our newly established skill in divisibility can lead us to a solution rather quickly. (Note that if you chose to buy 31 items, they would then cost $0.11, which does not conform with the original specifications of the problem.)

Here is another simple problem that can be further used as a warm-up as we continue our journey through some logical reasoning.

Consider a couple where the husband is older than the wife and the difference between the ages is $\frac{1}{11}$ of the sum of their ages. Furthermore, the wife's age is a reverse number of the husband's age. We now need to find their ages.

One approach might be to consider multiples of 11, since we know that the difference of the ages must be $\frac{1}{11}$ of the sum of their ages. One possibility would be that if the sum of their ages is 99, then the difference of their ages would be 9 years, so one possibility would be that their ages are 54 and 45, which satisfies all the conditions we were given at the outset. Just a little bit of logical reasoning guided us nicely to that answer.

We are now ready to embark on a problem which on the surface appears to be very simple and yet may be challenging to solve. However, the solution is so unexpected and simple to understand, which makes the problem quite entertaining.

## Don't Wine Over This Unexpected Result

The beauty of this proposed problem rests in the elegant solution offered later — as unexpected as it is — it almost makes the problem trivial and will truly entertain the audience. However, our conventional thinking patterns will likely cause a confusing haze over the problem. Don't allow your audience to despair. Encourage them to give it a genuine try. Or maybe even struggle a bit. Here is the problem.

> We have two one-gallon bottles. One contains a quart of red wine and the other, a quart of white wine. We take a tablespoonful of red wine and pour it into the white wine bottle. Then we take a tablespoon of this new mixture (white wine and red wine) and pour it into the bottle of red wine. Is there more red wine in the white wine bottle, or more white wine in the red wine bottle?

To solve the problem, we can figure this out in any of the usual ways — often referred to in the high school context as "mixture problems" — or we can use some clever logical reasoning. We begin as follows: With the first "transport" of wine there is only red wine on the tablespoon. On the second "transport" of wine, there is as much white wine on the spoon as there is red wine in the "white-wine bottle." This may require your audience to think a bit, but most should "get it" soon. Give them some time to reflect over the problem.

Now here comes the entertaining part: an unexpected solution! The simplest intelligible solution and the one that demonstrates a very powerful strategy is that of considering *extremes*. We use this kind of reasoning in everyday life when we resort to the option: "such and such would occur in a *worst-case scenario*, so we can decide to ...."

Let us now employ this strategy for the above problem and with it truly entertain and impress the audience. To do this, we will consider the tablespoonful quantity to be a bit larger. Clearly, the outcome of this problem is independent of the quantity transported, as long as it stays consistent throughout. So, we will use an *extremely* large quantity. We will let this quantity actually be the *entire* one quart. That is, following the instructions given in the problem statement, we will take the entire amount (one quart of red wine), and pour it into the white-wine bottle. This mixture is now 50% white wine and 50% red wine. We then pour one quart of this mixture back into the red-wine bottle. The mixture is now the same in both bottles. Therefore, there is as much white wine in the red-wine bottle as there is red wine in the white-wine bottle! The beauty of this problem lies in this solution procedure.

We can actually modify the procedure by considering another form of an extreme case, where the spoon doing the wine transporting has a zero quantity. In this case, the conclusion follows immediately: There is as much red wine in the white-wine bottle as there is white wine in the red-wine bottle, that is, zero! Carefully presented, this solution can be very significant in the way people approach future mathematics problems and even how they may analyze their everyday decision-making.

## A Simple Challenge

It is always entertaining to present what appears to be a very simple challenge to allow your audience a little bit of frustration and then the enjoyment of seeing how simple the solution could be. Let's take, for example, the geometric configuration shown in Figure 2.1. This configuration is comprised of 18 matchsticks and shows 8 triangles. The challenge here is to remove four of these matchsticks so that there are only 5 triangles remaining.

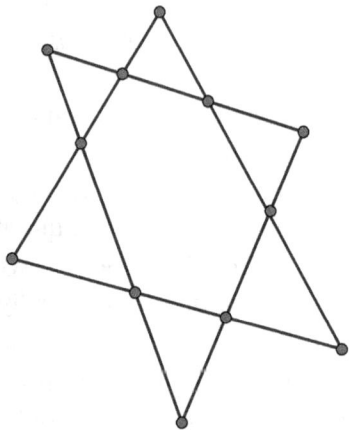

**Figure 2.1**

Once your audience has had a chance to attempt a solution, or perhaps reach a solution, you might want to expose a correct response to the challenge, which we show in Figure 2.2.

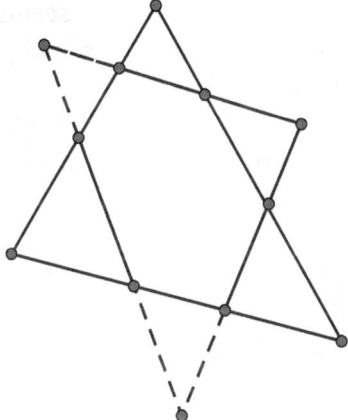

**Figure 2.2**

## A Counter-Intuitive Challenge

You can entertain your audience with a challenge that has been often played in restaurants and bars using toothpicks. Suppose you are presented with a collection of toothpicks arranged as shown in Figure 2.3, where each of the two rows and two columns contains 11 toothpicks.

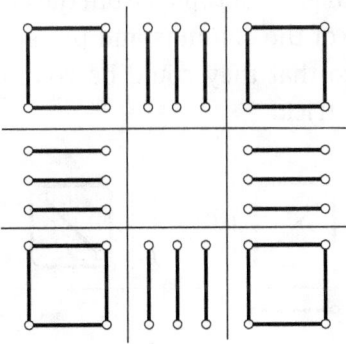

**Figure 2.3**

The challenge here is that your audience is asked to remove one toothpick from each row and column and still remain with 11 toothpicks in each row and column. This seems to be impossible, since we are actually *removing* toothpicks, and yet we are asked to keep the same number of toothpicks in each row and column, as before. Initially, this could perhaps look like the arrangement shown in Figure 2.4, but then there are not 11 toothpicks in each row and column as the challenge requested.

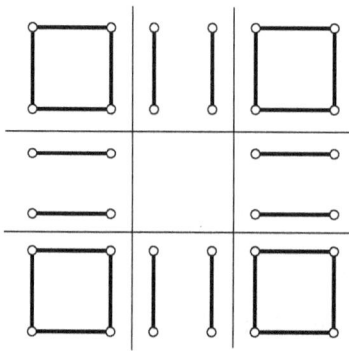

**Figure 2.4**

Your audience will then ask how can this possibly be done? After some contemplation, your audience will begin to see that if this can be done, they would have to count some toothpicks twice. In Figure 2.5, we see that we have taken a toothpick from the center portion of each of the rows and each of the columns and placed these toothpicks in the corner position so that they could be counted more than once. This is the crux of the trick!

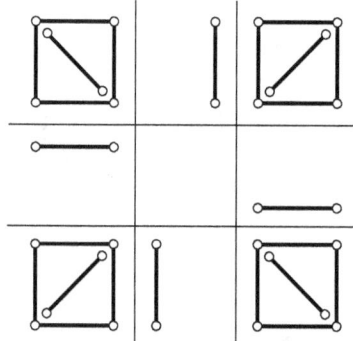

**Figure 2.5**

Thus, we have achieved our goal of having 11 toothpicks in each of the two rows and each of the two columns. Entertaining as it is, this is a skill that merits attention, so that one can go through life by analyzing situations in a more critical fashion.

Another counterintuitive example using matchsticks can be seen with the following challenge. Figure 2.6 shows four matchsticks the shape of a cup containing a ball. The challenge here is to move at most two of the matchsticks so that the ball is no longer in the cup.

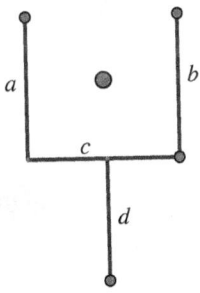

**Figure 2.6**

The moves here are rather deceptive. The first match that needs to be moved is match *c*, which needs to be slid to the right until the endpoint coincides with the endpoint of match *d*, and match *b* will have its endpoint in the middle of match *c*. We then move match *a* to close the cup as shown in Figure 2.7. Thus, we have moved only two matches namely matches *c* and *a*, and the ball is now outside of the cup. Might there be any other solutions?

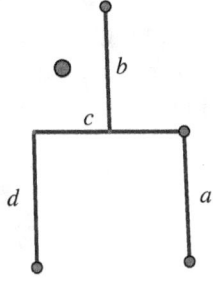

**Figure 2.7**

Admittedly this was not an easy task, but it does provide us with an illustration of logical thinking at work.

## Connecting the Dots

When asked to use 4 straight lines to connect the 9 dots in Figure 2.8, and doing so without lifting the pencil when drawing the lines, most people would go along the four sides of the square only to realize that the center dot has been omitted. The next step would be to include the center dot and then find that other dots have been omitted. For most people, there appears to be a psychological barrier to have the lines *extend* beyond the square. It is curious that most people won't "think out-of-the-box!"

Once again, here is the problem: Given 9 dots as arranged in Figure 2.8, without lifting the pencil, use 4 straight lines to connect all 9 dots.

**Figure 2.8**

After your audience feels a bit frustrated, you might enlighten them with the solution offered in Figure 2.9. Here the logical thinking requires that we stretch our preconceived notions of staying within the square as we see in the figure.

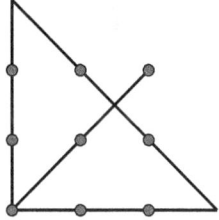

**Figure 2.9**

## Tooth-Pick Placement Corrections

We offer here an entertaining activity that can take place at the dinner table where you might have some toothpicks available. In each of the three cases, you are to place the toothpicks in such a way that the Roman numerals shown provide an incorrect equation. The challenge to your audience is to move exactly one toothpick and make the statement correct. In Figure 2.10, the left column shows the incorrect statement and the right column shows a correction by having moved only one toothpick. Note that in the second equation the X in the correction is a multiplication sign and in the third equation the square arrangement represents a zero.

| Incorrect statements | Correction with one toothpick moved |
|---|---|
| $\lvert\lvert - \lvert\lvert\lvert = \lvert$ | $\lvert\lvert = \lvert\lvert\lvert\, \bar{\ }\, \lvert$ |
| $\mathsf{X} - \lvert\lvert = \lvert\lvert$ | $\lvert \mathsf{X} \quad \lvert\lvert = \lvert\lvert$ |
| $\lvert\lvert + \square = \lvert$ | $\lvert\lvert - \square = \lvert\lvert$ |

**Figure 2.10**

## Move Circles and Reduce Rows

In Figure 2.11, we show an arrangement of 12 dots in six straight rows. Our challenge here is to see how we can move 4 of these dots to other positions so that we are left with 7 rows and still remaining with 4 dots in each row.

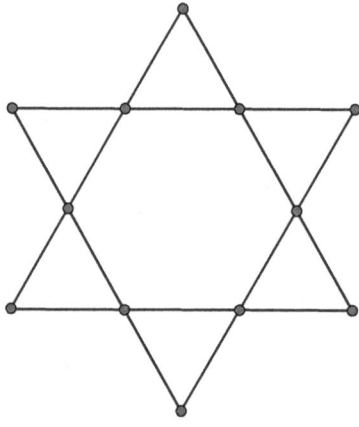

**Figure 2.11**

Your audience will initially be somewhat frustrated and yet at the same time locked into maintaining the star-shape shown here. The solution will be to break out of the "mold." Your aim as entertainer is to lead your audience to the solution we show in Figure 2.12.

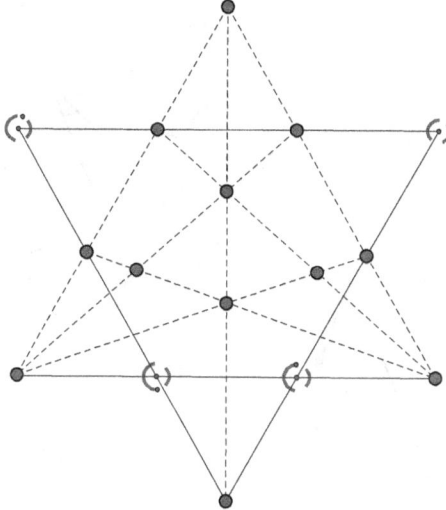

**Figure 2.12**

There you can see the 4 dots that we removed and placed collinearly with the existing dots so that we now have all 12 dots situated so that only 7 lines are needed. Although this may require some "thinking out-of-the-box," it will, nonetheless, prove to be entertaining as well as instructional.

## Arranging Dots (or Plants)

Here the problem is to arrange 21 dots so that they form 12 straight rows with 5 dots in each row. To make this more entertaining, you might choose to present this in the form of a planting problem, where an artistic arrangement is planned and one only has 21 plants available and yet decides to make the pattern as described above. The solution, which we show in Figure 2.13, begins with an octagon in the central part of the figure and then the various sides of the octagon are extended to their common intersections. When you count the dots, you will find that there are 5 in each row and there are 12 such rows, which is what was requested in the original challenge.

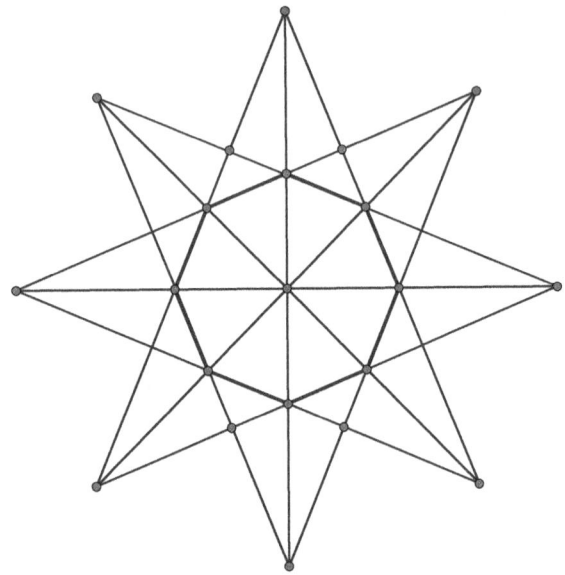

**Figure 2.13**

## Placing Dots Appropriately

Here you can entertain your audience in a rather simple way by either using dots as we show in Figure 2.14 or simply marking the dots on a blank sheet of paper as shown there. The challenge here is to move 4 of the 10 dots shown and placed them in such a way that the resulting arrangement will allow 5 straight lines connecting all the dots, where each line contains 4 dots.

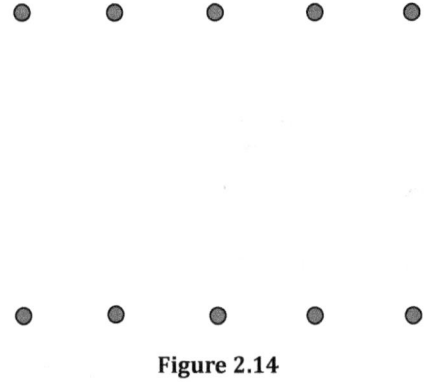

**Figure 2.14**

There are many ways in which this challenge can be met. We will show a few of them here, but you might challenge your audience to come up with other arrangements. One solution can be found in Figure 2.15.

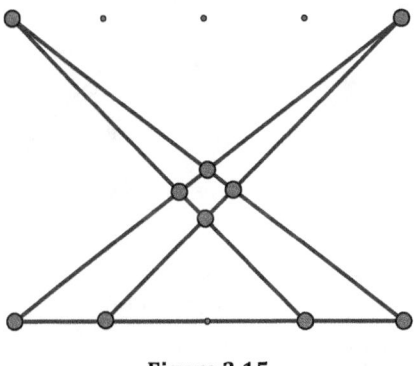

**Figure 2.15**

Another possible solution can be found in Figure 2.16.

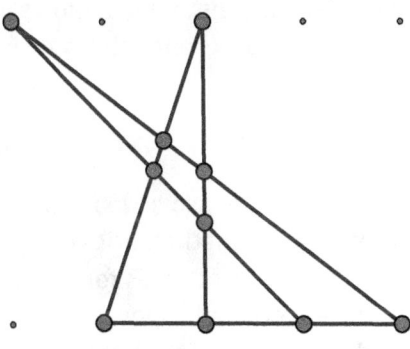

**Figure 2.16**

In Figure 2.17, we offer yet another version that should allow the audience to create other versions as well once they see a pattern.

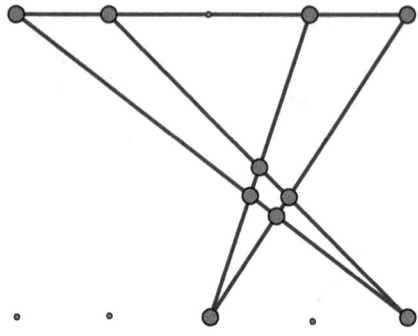

**Figure 2.17**

To summarize the procedure in general terms, one can see that you can select 3 dots from the bottom row in 10 different ways and 1 dot from the top row in 5 different ways which would allow for 50 different combinations to be created. If we flip that around, we would have another 50 ways of getting this solution. Therefore, we would have 100 ways thus far. However, we can select which dots are going to be eliminated in 24 different ways, which allows us then to do this problem with $24 \times 100 = 2400$ different solutions.

## The Matchstick Plan

Presenting a problem in story form as we do here might give the audience something to think about as mathematics can also solve practical problems. In planning out a series of cages for animals, a farmer used matchsticks to map out the series of cages, as we see in Figure 2.18. In doing so, he used 13 matches and was able to produce 6 cages of equal size. He used matchsticks to represent fencing units.

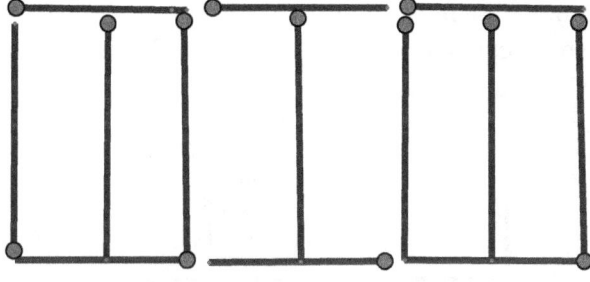

**Figure 2.18**

As he was about to begin the actual construction according to the match stick pattern shown in Figure 2.18, he realized that he was lacking 1 fencing unit, and only had therefore, 12 fencing units, which he now will use again to try to create 6 cages of equal size. After you had your audience try to create 6 equal cages with only 12 matchsticks, they will be quite surprised when you show them how this can be done, as we show in Figure 2.19.

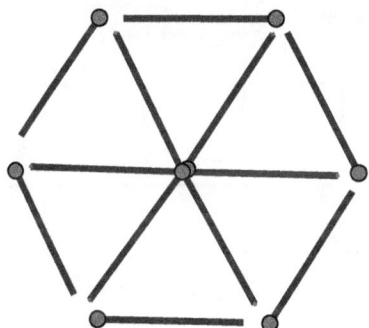

**Figure 2.19**

## The Missing Square

In this challenge, your audience will have to "think out-of-the-box." Yet, such activities can surely be enriching and also instructional — although they would be a good way of training unusual thinking. Here is the problem: Removing a square by just moving 2 sticks rather than 4, can present a curious logical problem. In Figure 2.20, we show 5 squares. How can we eliminate one of the squares so that only 4 squares remain by moving only 2 sticks?

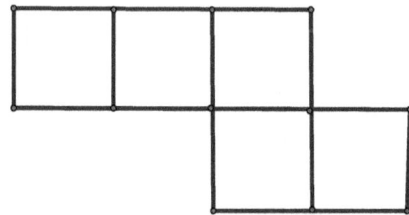

**Figure 2.20**

After the audience has had time to ponder this dilemma, it would be wise to show a solution which should certainly get a "gee-whiz" reaction. Figure 2.21 shows which two sticks can be moved to leave only 4 squares.

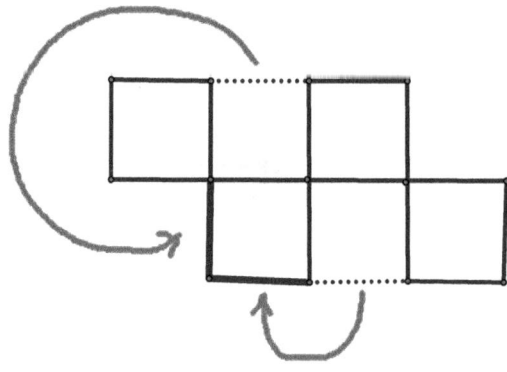

**Figure 2.21**

## Creating a Pentagon

Once again, the audience is given a logistical problem to solve. We show in Figure 2.22 two triangles that are overlapping and forming a third triangle.

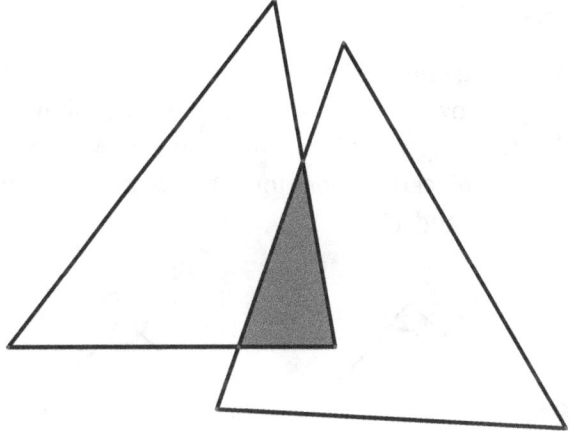

**Figure 2.22**

The challenge here is to rearrange the two large triangles so that their overlap creates a Pentagon. Ample time should be provided for the audience to experiment with various attempts. After sufficient time has been provided, you may show the solution in Figure 2.23.

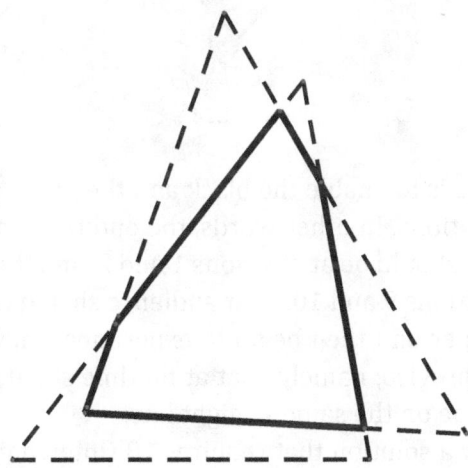

**Figure 2.23**

## Strategic Moves

There are times when one can be entertained by the challenge of making some strategic moves as we will show with the following strategy game. Consider the diagram shown in Figure 2.24, where we begin with 2 black chips placed at positions 1 and 2, and 2 white chips placed at positions 9 and 10.

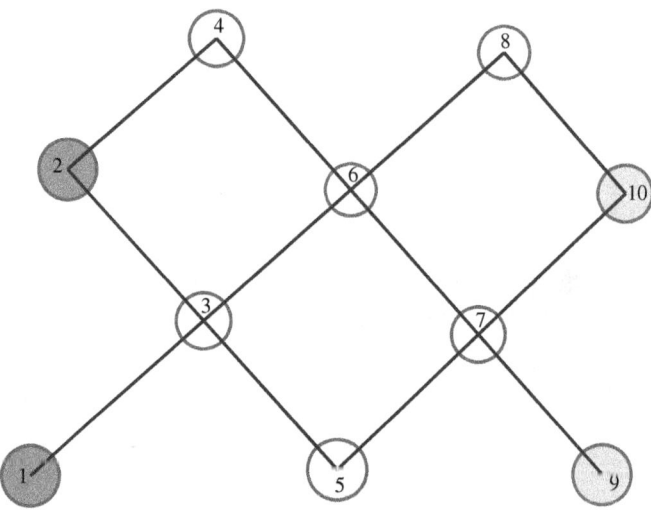

**Figure 2.24**

The challenge is to enable the black and the white chips to interchange their positions. In other words, the end result should be that the 2 white chips should be at positions 1 and 2, and the 2 black chips should be at positions 9 and 10. Your audience should copy the figure on a sheet of paper and then begin to experiment moving the chips along, with one proviso, namely, that at no time should a white chip and a black chip be on the same straight line.

We offer now a solution that requires 18 (interchange) moves as follows:

$$2 - 3, 9 - 4, 10 - 7, 3 - 8, 4 - 2, 7 - 5, 8 - 6, 5 - 10, 6 - 9,$$
$$2 - 5, 1 - 6, 6 - 4, 5 - 3, 10 - 8, 4 - 7, 3 - 2, 8 - 1, 7 - 10$$

# Think Before Counting

We can entertain the audience with a problem situation that seems so simple that we plunge right in without first thinking about a strategy to use. This impetuous beginning for the solution often leads to a less elegant solution than one that results from a bit of forethought. Here are two examples of simple problems that can entertain an audience by the overlooked simplicity of the question posed.

*Find all pairs of prime numbers whose sum equals 999.*

Some members of your audience will begin by taking a list of prime numbers and try various pairs to see if they can obtain 999 for a sum. This is obviously very tedious, as well as time consuming, and you would never be quite certain that you had considered all the prime-number pairs. Here is where you entertain the audience by showing them how some logical reasoning makes this problem trivial. In order to obtain an odd sum (in this case 999) for two numbers (prime or otherwise), exactly one of the numbers must be even. Since there is only one *even* prime, namely, 2, there can be only one pair of primes whose sum is 999, and that pair is 2 and 997. An interesting reaction can be expected from an audience that has spent some time experimenting by taking the sum of a variety pairs of primes.

A second problem where pre-planning, or some orderly thinking makes sense is as follows:

*A palindrome is a number that reads the same forwards and back-wards, such as 747 or 1991. How many palindromes are there between 1 and 1000 inclusive?*

The traditional approach to this problem would be to attempt to write out all the numbers between 1 and 1000, and then see which ones are palindromes. However, this is a cumbersome and time-consuming task at best, and one could still easily omit some of them.

Let's see if we can look for a pattern to solve the problem in a more elegant and impressive fashion.

| Range | Number of palindromes | Total number |
|-------|----------------------|--------------|
| 1–9 | 9 | 9 |
| 10–99 | 9 | 18 |
| 100–199 | 10 | 28 |
| 200–299 | 10 | 38 |
| 300–399 | 10 | 48 |
| ⋮ | ⋮ | ⋮ |
| 900–999 | 10 | 108 |

Curiously, a pattern has evolved, namely that there are exactly 10 palindromes in each group of 100 numbers (after 99). Thus, there will be 9 sets of 10, or 90, plus the 18 from numbers 1 to 99, for a total of 108 palindromes between 1 and 1000. Once again, the audience will awed by the simple solution presented.

Another solution to this problem would involve organizing the data in more favorable way. Consider all the single digit numbers (self-palindromes). There are 9 such numbers. There are also 9 two-digit palindromes. The three-digit palindromes have 9 possible "outside digits" and 10 possible "middle digits," so there are 90 of these. In total, there are 108 palindromes between 1 and 1000, inclusive. The motto here should be: Think first, then begin a solution!

## Think Before Starting to Solve This

Here is one that is very difficult if you don't see the trick, and can be trivial if you do see the clever way to solve this problem.

> *If the sum of 2 numbers is 2 and the product of the same 2 numbers is 3, find the sum of the reciprocals of these 2 numbers.*

Most readers would immediately revert to algebra and set up the following two equations: $x + y = 2$, and $xy = 3$. Typical algebra training leads us to prepare to solve these two linear equations simultaneously. That would have us change the first equation to read $y = 2 - x$, and then substitute this value for $y$ into the second equation to get: $x(2-x)=3$, which leads to the quadratic equation $x^2-2x+3=0$. Using the quadratic formula[1] to solve this equation, we find that $x=1\pm i\sqrt{2}$. We would then need to find the value of $y$. Then we would have to take the reciprocals, and then to add them to get the required answer, which is a rather cumbersome method of solution. The curious part of this original problem is that it can be solved very simply, if we focus on what we have been asked to find, and not be distracted by finding the values of $x$ and $y$. We have been asked to find the sum of the reciprocals, not necessarily the values of $x$ and $y$. That is, we actually only need to find $\frac{1}{x}+\frac{1}{y}$. So let's find the sum of the reciprocals: $\frac{1}{x}+\frac{1}{y}=\frac{x+y}{xy}$, which essentially gives us the answer, since we know both the numerator and denominator from the given information: $x+y=2$, and $xy=3$. Hence, $\frac{1}{x}+\frac{1}{y}=\frac{x+y}{xy}=\frac{2}{3}$, and our problem is solved. Notice, by working backwards we achieved a very elegant solution to a problem that otherwise would have been rather complicated to solve.

## Making a Closed Chain

As we have mentioned earlier, there are times when logical thinking means "thinking out-of-the-box." This is the case here where we are presented with 4 pieces of chain, consisting of 3 links each (see Figure 2.25). The challenge to the audience is to show how these 4 pieces of chain could be made into a circular chain of 12 links by opening and closing, *at most*, 3 links.

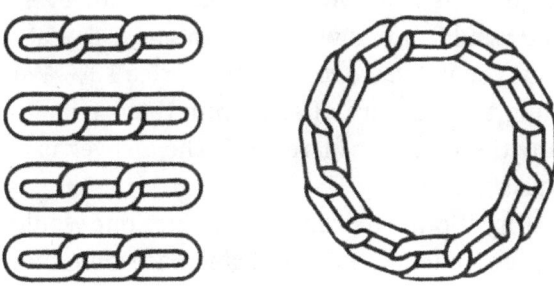

**Figure 2.25**

Typically, a first attempt at a solution involves opening the end link of one chain piece and then joining it to the second chain piece to form a 6-link chain; then opening and closing a link in the third chain piece and joining it to the 6-link chain to form a 9-link chain. By opening and closing a link in the fourth chain piece and joining it to the 9-link chain, we get a 12-link chain piece, which is *not* a circle, and so we have done 3 chain link open–close and still have not obtained a circular chain. Thus, this typical attempt ends unsuccessfully. Most attempts usually involve other combinations of open/closing a link of each of various chain pieces to try to join them together to get the desired result, but this approach will not work.

Let's look at this from another point of view. Instead of continually trying to open and close *one* link of each chain piece, a different point of view would involve opening *all* the links in one chain and using those links to connect the remaining three chain pieces together, and into the required circle chain. This quickly gives the successful solution, and should prove to be an interesting "eye-opener" for the audience.

## The Worthless Increase

Here is a situation that you can use to perhaps "upset" the audience — of course, in a favorable sense — with the unexpected result. Present your audience with the following situation. Suppose you have an item you want to sell to a person and are willing to give him a 10% discount, if he buys it from you immediately. However, the person decides that he would rather wait until the next day at which point you tell him that you will have to raise that price 10%. Would the person be assuming that you are offering him the original price for the item before any discounts or increases? The answer is a resounding (and very surprisingly): No!

This little story is quite disconcerting, since one would expect that with the same percent increase and decrease one should be back to where one started. This is intuitive thinking, but it is wrong! To explain this to your friend you might have your friend choose a specific amount of money to begin with. Suppose we begin with $100. Calculate a 10% decrease and then the 10% increase. Using a $100 basis, we first calculate a 10% decrease to get $90. Then the 10% increase, which is $9, and therefore, yields $99, which is $1 less than the original amount. Therefore, the person benefits nevertheless by having waited to the next day to buy your item.

You may wonder whether the result would have been different if we had first calculated the 10% increase on the $100 to get $110. Then take a 10% decrease of this $110, which is $11, to get $99 — the same as before. So, clearly the order makes no difference.

A similar situation, one that is deceptively misleading, can be faced by a gambler. Consider the following situation. Offer your audience the following challenge. You are offered a chance to play a game, where the rules are as follows: There are 100 cards, face down. Of those cards, 55 cards say "win" and 45 of the cards say, "lose." You begin with $10,000. You must bet one-half of your money on each card turned over, and you either win or lose that amount based on what the card says. At the end of the game, all cards have been turned over. How much money do you have at the end of the game?

The same principle as above applies here. It is obvious that you will win 10 times more than you will lose, so it appears that you will end with more than $10,000. What is obvious is often wrong, and this is a good example. Let's say that you win on the first card; you now have $15,000. Then you lose on the second card; you now have $7500. If you had first lost and then won, you would still have $7500. So, every time you win one and lose one, you lose $\frac{1}{4}$ of your money. So, you end up with $10,000 \cdot \left(\frac{3}{4}\right)^{45} \cdot \left(\frac{3}{2}\right)^{10}$. This results in $1.38 when rounded off. Surprised? With this knowledge, a mean trickster can surely take advantage of a "friend."

## A Little Reasoning

Sometimes problems seem awfully simple, and yet, to get the correct solution a bit more reasoning is required. Consider for example, the students are playing cards where the winner of each game wins

1 dollar. At the end of the night, one player has won 5 games and the other player has won 5 dollars. The question is how many games had to have been played to get this result?

Through some logical reasoning, we would come to the conclusion that one student must have won 5 games in order to win 5 dollars. The second student would have had to have won 5 games so they would be even and then won another 5 games, which would leave him with 5 dollars. By counting of these games, we arrive at a total of 15 games played.

## Coin Entertainments

There are some mathematics entertainments that require some logical thinking — and can be explained through algebra or other traditional aspects of mathematics. However, above all they are easy to follow and have truly unexpected results which ought to be genuinely entertaining. As an illustration, one such can be done with coins and will show your audience how some clever reasoning along with very elementary algebraic knowledge will help sort out this unexpected result. Although this problem was offered in the introduction, we are providing it again since it fits in the chapter and is worth repeating.

Suppose your friend is seated at a table in a dark room. On the table, there are 12 pennies, 5 of which are heads up and 7 are tails up. She knows where the coins are, so she can mix them by sliding around the coins, but because the room is dark, she will not know if the coins that she is touching were originally heads up or tails up. You now ask her to separate the coins into two piles of 5 and 7 coins and then flip all the coins in the 5-coin group. To everyone's amazement, when the lights are turned on there will be an equal number of heads in each of the two piles. Your friend's first reaction is "you must be kidding!" How can anyone do this task without seeing which coins are heads up or tails up? The solution will surely enlighten the friend and at the same time illustrate how algebraic symbols can help understanding.

Let's now look at the explanation of this surprise result. This is where a clever (yet incredibly simple) use of algebra will be the key to explaining the unexpected outcome. Let's "cut to the chase." The 12 coins have 5 with heads up and 7 with tails up. Without being able to look at the coins, she separated the coins into two piles, of 5 and 7 coins each. Then she flipped over the coins in the smaller pile of 5 coins. Then both piles had the same number of heads!

Well, this is where a little algebra helps us understand what was actually done. Let's say that when she separated the coins in the dark room, $h$ heads will end up in the 7-coin pile. Then the other pile, the 5-coin pile, will have $5 - h$ heads. To get the number of tails in the 5-coin pile, we subtract the number of heads $(5 - h)$ from the total number of coins in the pile, 5, to get: $5 - (5 - h) = h$ tails.

| 5-coin pile | 7-coin pile |
|---|---|
| $5 - h$ heads | $h$ heads |
| $5 - (5 - h)$ tails $= h$ tails | |

When she flips all the coins in the smaller pile (the 5-coin pile), the $(5 - h)$ heads become tails and the $h$ tails become heads. Now each pile contains $h$ heads!

**The piles after flipping the coins in the smaller pile**

| 5-coin pile | 7-coin pile |
|---|---|
| $5 - h$ tails | $h$ heads |
| $h$ heads | |

This absolutely surprising result will show you how the simplest algebra can explain an entertaining aspect of mathematics.

With these brief examples, we hope to have whet your appetite for the many unexpected and counterintuitive entertainments that the field of mathematics can offer. Let's continue a journey through more of such recreational aspects of mathematics.

## Clever Guesses

If you want to impress your audience you might want to present a nice logical problem with a somewhat unexpected solution. Consider that you have 3 boxes each containing 2 balls. One box has 2 black balls, a second box has 2 white balls and one box has a black and a white ball. We show these 3 boxes in Figure 2.26, however, the boxes are all mislabeled. In other words, none of the boxes describes its contents properly.

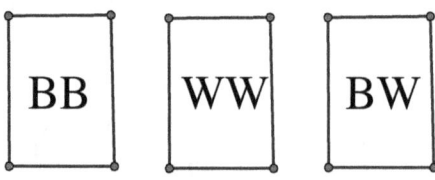

**Figure 2.26**

The question posed is how can we determine the true contents of each box by taking out only one ball at a time from a box in the most efficient way? That is, we seek the least number of samplings to determine the contents of the boxes. The beauty of this solution is that we can determine the contents of each of the boxes by taking *only one ball from one box* in a clever fashion. Keep in mind that all the labels on the boxes you see in Figure 2.26 are incorrectly placed.

Here is the best solution, where we will take one ball from box labeled BW and then use the following logic to determine the boxes' contents. If the ball drawn is black, then you would know that the other ball cannot be white, since the BW was incorrectly labeled. So, that must be the box with the 2 black balls.

We know that the box labeled WW, cannot contain 2 white balls. Therefore, since it cannot be the box that contains 2 black balls, which we have already identified, it must be the box that has a black and white ball. It, therefore, remains that the box labeled BW must contain the 2 white balls. This will surely give your audience something to think about, especially the power of logical thinking.

# Alternate Ways of Thinking

Of the many strategies that exist and are particularly useful solving some mathematical problems, one of which is the strategy of approaching the problem from a different point of view. This is one that allows us to avoid "running into the wall" — namely avoiding frustration. Perhaps one of the classic examples — because of its simplicity and dramatic difference in solution methods — is the following. This is an example where the common method leads to a correct answer but is cumbersome and can often provide an opportunity for some arithmetic mistakes. The problem is as follows:

> At a school with 25 classes, each of these sets up a basketball team to compete in a school-wide tournament. In this tournament a team that loses one game is immediately eliminated. The school only has one gymnasium, and the principal of the school would like to know how many games will be played in this gymnasium in order to get a winner.

The typical solution to this problem could be useful. To simulate the actual tournament, we begin with 12 randomly selected teams playing a second group of 12 teams — with one team drawing a bye — that is passing up a game. This would then continue with the winning teams playing each other as shown here.

- Any **12 teams** vs. any other **12 teams,** which leaves **12 winning teams** in the tournament.
- **6 winners** vs. **6 other winners,** which leaves **6 winning teams** in the tournament.
- **3 winners** vs. **3 other winners,** which leaves **3 winning teams** in the tournament.
- **3 winners + 1 team** (which drew a bye) = **4 teams.**
- **2** remaining **teams** vs. **2** remaining **teams,** which leaves **2 winning teams** in the tournament.
- **1 team** vs. **1 team** to get a **champion!**

Now counting up the number of games that have been played we obtain:

| Teams playing | Games played | Winners |
|:---:|:---:|:---:|
| 24 | **12** | 12 |
| 12 | **6** | 6 |
| 6 | **3** | 3 |
| 3 + 1 bye = 4 | **2** | 2 |
| 2 | **1** | 1 |

The total number of games played is $12 + 6 + 3 + 2 + 1 = 24$.

This seems like a perfectly reasonable method of solution, and certainly a correct one. Approaching this problem from a different point of view would be vastly easier, that is, by considering the losers rather than the winners, which is what we did in the previous solution. In that case, we ask ourselves, how many losers must there have been in this competition in order to get one champion? Clearly, starting with the 25 original teams there had to be 24 losers. To get 24 losers, there needed to be 24 games played, and with that the problem is solved. Looking at the problem from an alternative point of view is a curious approach that can be useful in a variety of contexts.

Another alternative point of view would be to consider these 25 teams with one of them — only for our purposes — considered to be a professional basketball team that would be guaranteed to win the tournament. Each of the remaining 24 teams would be playing the professional team only to lose. Once again, we see that 24 games are required to get a champion. This should demonstrate for your audience the power of this problem-solving technique.

## Working with Infinity

Too often folks in your audience may be taking the concept of infinity for granted. However, when asked to compare the size of the infinite set of natural numbers {1, 2, 3, 4, 5,....} to the infinite set of even numbers {2, 4, 6, 8, ....}, they would probably have a difficult time to accept the fact that they are of equal size. The ordinary person sees this as illogical, since all the odd numbers are missing from the second set, how can the 2 sets of the same number of elements be of equal size? The best explanation is that for every number in the set of natural numbers, there is a partner element in the set of even numbers; so that there is a one-to-one correspondence between the 2 sets, making them equal in size. This is only true because they are *infinite* in size.

Let's see how this concept of infinity can help us with a geometric problem that involves infinity.

In Figure 2.27, we have an isosceles triangle with an infinite series of circles, each of which is tangent to the 2 sides of the isosceles triangle and to the adjacent circles. The sides of the isosceles triangle are 13, 13, and 10. We need to find the sum of the circumferences of these circles.

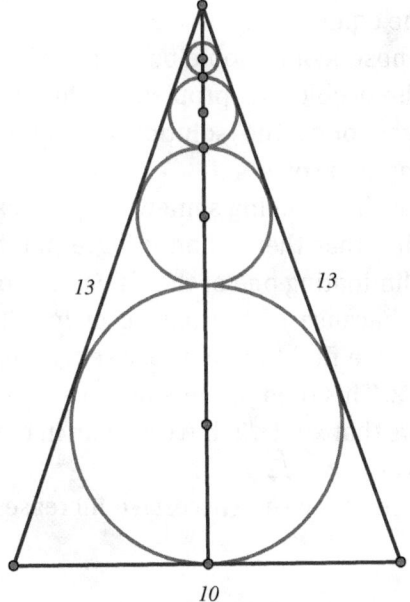

**Figure 2.27**

As tedious as it sounds, a common approach here would be to find the circumference of each of the circles and then take their sum. This would be a very complicated computation, but done carefully, it could lead to a correct solution.

A much more elegant approach would be to consider the special aspects of infinity and consider this problem from another point of view. Using the Pythagorean theorem, we find that the altitude of this isosceles triangle has length 12. We also will notice that the sum of the diameters is equal to the altitude of the isosceles triangle, since there is an infinite number of circles. The circumference of one circle is $\pi$ times its diameter. Therefore, the sum of the circumferences of the infinite number of circles is the sum of the diameters times $\pi$, which is $12\pi$. Here again, your audience can get an appreciation of what infinity brings to the mathematics table.

Now that the audience is getting somewhat more comfortable with the concept of infinity you might have them try to find the value of $x$ that satisfies the equation: $x^{x^{x^{x^{\cdots}}}} = 2$.

At first glance, most people would be overwhelmed, and not know how to approach the problem. A problem of this sort could be quite frustrating. However, once the solution is exposed the problem becomes rather simple to solve.

We could look at this as being somewhat of an extreme situation. We begin by noticing that there is an infinite number of $x$'s in this tower of powers. Eliminating one of the $x$'s would not have any effect on the end result, because of the nature of infinity. Therefore, by removing the first $x$, we find that all those remaining in the tower of $x$'s must also equal 2. This then permits us to rewrite this equation as $x^2 = 2$. It then follows that $x = \pm\sqrt{2}$. If we remain in the set of real numbers, then the answer is $x = \sqrt{2}$.

Below you can see how the successive increases get ever closer to 2.

$$\sqrt{2} = 1.414213562\ldots$$

$$\sqrt{2}^{\sqrt{2}} = 1.632526919\ldots$$

$$\sqrt{2}^{\sqrt{2}^{\sqrt{2}^{\sqrt{2}}}} = 1.760839555...$$

$$\sqrt{2}^{\sqrt{2}^{\sqrt{2}^{\sqrt{2}^{\sqrt{2}}}}} = 1.840910869...$$

$$\sqrt{2}^{\sqrt{2}^{\sqrt{2}^{\sqrt{2}^{\sqrt{2}^{\sqrt{2}}}}}} = 1.892712696...$$

$$\sqrt{2}^{\sqrt{2}^{\sqrt{2}^{\sqrt{2}^{\sqrt{2}^{\sqrt{2}^{\sqrt{2}}}}}}} = 1.926999701...$$

$$\vdots$$

And so, we have a surprisingly simple solution to a very complicated looking problem.

Just another example of how you can simplify problems involving infinity. Instead of finding the value of the following infinite nest of radicals: $\sqrt{2+\sqrt{2+\sqrt{2+\sqrt{2+\sqrt{2+\sqrt{2+\cdots}}}}}}$

Once again, recognizing the effect of infinity, we can let this be equal to $x$, so that $x = \sqrt{2+\sqrt{2+\sqrt{2+\sqrt{2+\sqrt{2+\sqrt{2+\cdots}}}}}}$ and then squaring both sides, we have $x^2 = 2+\sqrt{2+\sqrt{2+\sqrt{2+\sqrt{2+\sqrt{2+\sqrt{2+\cdots}}}}}}$ . The infinite nest of radicals does not change its value when we remove one unit, so that this equation can be written as $x^2 = 2+x$, which in the form of $x^2 - x - 2 = 0$ can be easily solved, since we have $(x-2)(x+1)=0$, where the positive answer would be $x = 2$. These examples should give your audience plenty to think about as to how we can handle infinity elegantly.

## A Square Arrangement of 9 Numbers

The first 9 numbers also lend themselves to an arrangement which also produces quite a bit of entertaining oddities. When numbers are arranged in a square and provide the same sum for each row, column, or diagonal we call such a square a *magic square.* One such arrangement can be seen in Figure 2.28.

| 4 | 9 | 2 |
|---|---|---|
| 3 | 5 | 7 |
| 8 | 1 | 6 |

**Figure 2.28**

Once you show this magic square to your audience, they will want to test to make sure that the sum of each row, column, and diagonal is the same, which in this case 15. To further entertain them with this magic square, you might ask them to see what the relationship among the numbers in each row, column, and diagonal is. They should notice that the differences between the numbers is a constant. For example, in the diagonal containing 8, 5, and 2, the numbers differ by 3; while in the middle row, 3, 5, and 7, the numbers differ by 2. And so, they will find a common difference in each of these 8 triplets of numbers.

There are lots more peculiarities of this magic square that could be entertaining, for example, if we take the squares of each of the numbers in the magic square, we will find that the sums of the squares in the first and third columns are equal. This can be seen as $4^2 + 3^2 + 8^2 = 16 + 9 + 64 = 89$, and $2^2 + 7^2 + 6^2 = 4 + 49 + 36 = 89$.

Similarly, the sums of the squares of the first and third rows are also equal, as we can see with: $4^2 + 9^2 + 2^2 = 16 + 81 + 4 = 101$ and $8^2 + 1^2 + 6^2 = 64 + 1 + 36 = 101$.

To further provide some entertainment, as if this not yet enough for the first 9 numbers, consider the sum of the squares in the middle column and the middle row:

The sum of the squares in the middle column is $9^2 + 5^2 + 1^2 = 81 + 25 + 1 = 107$, and the sum of the squares of the middle row is $32 + 52 + 72 = 9 + 25 + 49 = 83$.

Now this is a little bit tricky, if we take the sums of the squares of the columns, 89, 107, 89, we find that the difference between the middle column and the two side columns is 18.

When we consider the sums of the squares of the three rows, 101, 83, 101, once again the difference between the sums of the squares of the middle row and the two side rows is 18. Frankly, this may be a little bit unexpected and may be difficult to find without some guidance.

As if this weren't enough, we still have another unusual aspect with which we can entertain the audience, that is, if we consider the triplets in each of the three rows as a number and take their squares to get $492^2 + 357^2 + 816^2 = 242{,}064 + 127{,}449 + 665{,}856 = 1{,}035{,}369$. We find that the sum is the same as the sum of the squares of the numbers with the digits in reverse, as we can see with $294^2 + 753^2 + 618^2 = 86{,}436 + 567{,}009 + 381{,}924 = 1{,}035{,}369$. The same pattern can be done with the columns, where $438^2 + 951^2 + 276^2 = 191{,}844 + 904{,}401 + 76{,}176 = 1{,}172{,}421$, and now reversing the numbers and finding the sum of their squares to get $834^2 + 159^2 + 672^2 = 695{,}556 + 25{,}281 + 451{,}584 = 1{,}172{,}421$. This is truly amazing and should be so emphasized when presenting it to others.

Furthermore, when we consider the sums of the numbers considered as a three-digit number of the diagonals and their reversals, we find that they are equal, as $456 + 654 = 1110$, and $852 + 258 = 1110$.

While we are still on the topic of a $3 \times 3$ magic square, we offer here for entertainment a most unusual magic square shown in Figure 2.29, where rather than addition, we *use multiplication* to show that the rows, columns, and diagonals all have the same product. By the way, this is the smallest possible number for a product (216) for any such $3 \times 3$ multiplication magic square, where all the cells have different numbers.

| 3 | 36 | 2 |
|---|----|---|
| 4 | 6 | 9 |
| 18 | 1 | 12 |

**Figure 2.29**

Suppose we now interchange the four corner cells with their opposite corners so that we get the square arrangement shown in

Figure 2.30. This square arrangement, which probably cannot be considered a magic square anymore, carries an unusual relationship. If you take the end numbers on any row, column, or diagonal and multiply them and then divide them by the center cell, the relationship will be that they are all equal, such as, for example, $12 \times 18 \div 36 = 6$ and $4 \times 9 \div 6 = 6$.

| 12 | 36 | 18 |
|----|----|----|
| 4  | 6  | 9  |
| 2  | 1  | 3  |

**Figure 2.30**

As we have been saying throughout, there is no shortage of surprises when it comes to mathematical surprises. Let us once again consider the original $3 \times 3$ magic square, which we considered earlier, and for convenience, we will once again show it as Figure 2.31.

| 4 | 9 | 2 |
|---|---|---|
| 3 | 5 | 7 |
| 8 | 1 | 6 |

**Figure 2.31**

We now interchange the four corner numbers with their opposites, as we did earlier with the conversion of the multiplication square to the division square, so that we get the following square arrangement, shown in Figure 2.32. This time we find that when we add the end numbers in any row, column, and diagonal and then subtract the center square number, the result will always be the same, namely, 5, for example, $6 + 8 - 9 = 5$ and $8 + 4 - 7 = 5$.

| 6 | 9 | 8 |
|---|---|---|
| 3 | 5 | 7 |
| 2 | 1 | 4 |

**Figure 2.32**

Were we to satisfy ourselves with only a partial magic square, where only the rows and columns have the same product, then we would have a magic square shown in Figure 2.33.

| 10 | 4 | 3 |
|----|----|---|
| 12 | 2 | 5 |
| 1 | 15 | 8 |

**Figure 2.33**

There is much more entertainment that can be done with magic squares, both with multiplication and with addition. Ambitious readers may want to try to construct others, as well as learning a procedure on how to construct them. We will consider the construction of proper magic squares later this chapter.

To add to the entertainment of magic squares consider the following magic square shown in Figure 2.34, where all the rows, columns, and diagonals have not only the same sum but also the same product. The smallest such magic square is an 8 × 8 magic square that was originally constructed by American mathematics teacher Walter W. Horner in 1955. The common sum of the rows, columns, and diagonals is 840 and the common product of these is 2,058,068,231,856,000. Of course, we will have to take Mr. Horner's word for it unless there is an ambitious member of the audience willing to do the product of each row, column, and diagonal to test the products.

| 162 | 207 | 51 | 26 | 133 | 120 | 116 | 25 |
| --- | --- | --- | --- | --- | --- | --- | --- |
| 105 | 152 | 100 | 29 | 138 | 243 | 39 | 34 |
| 92 | 27 | 91 | 136 | 45 | 38 | 150 | 261 |
| 57 | 30 | 174 | 225 | 108 | 23 | 119 | 104 |
| 58 | 75 | 171 | 90 | 17 | 52 | 216 | 161 |
| 13 | 68 | 184 | 189 | 50 | 87 | 135 | 114 |
| 200 | 203 | 15 | 76 | 117 | 102 | 46 | 81 |
| 153 | 78 | 54 | 69 | 232 | 175 | 19 | 60 |

**Figure 2.34**

# Non-magic Squares

Before we leave the topic of 3 × 3 square arrangements of numbers, we can challenge the audience to find square arrangements were **no** row, column, or diagonal has the same sum using the digits 1 to 9. There are many ways to create such a non-magic square. There are two rather attractive ones where the numbers appear in a rather spiraled fashion — going clockwise beginning at the upper left in Figure 2.35, and counterclockwise beginning at the center as shown in Figure 2.36.

| 1 | 2 | 3 |
| --- | --- | --- |
| 8 | 9 | 4 |
| 7 | 6 | 5 |

**Figure 2.35**

| 9 | 8 | 7 |
|---|---|---|
| 2 | 1 | 6 |
| 3 | 4 | 5 |

**Figure 2.36**

Your audience may wish to try to construct other non-magic squares, such as the one shown in Figure 2.37.

| 4 | 3 | 5 |
|---|---|---|
| 1 | 9 | 7 |
| 8 | 6 | 2 |

**Figure 2.37**

While it is challenging to create magic squares, it can also be challenging to create non-magic squares and this, too, can be entertaining.

## Indian Magic Square

Returning now to the square arrangements of numbers, usually referred to as *magic squares*, that have been a source of entertainment for centuries. Magic squares can be found scattered in ancient times; however, we will introduce a magic square that appeared in India in the tenth century, and was called Chautisa Yantra and is found in the Parshvanath Jain temple in Khajuraho, India (see Figure 2.38).

**Figure 2.38**

Perhaps it is appropriate that we begin with this Indian magic square, since we truly know that the origin of our numeral system in the Western world stems from India, as we mentioned earlier in Chapter 1. Recall that these numerals were first mentioned in Europe in the year 1202, in the first sentence of the introduction in Fibonacci's book *Liber Abaci*. This Indian 4 × 4 — magic square is shown in Figure 2.39, where the sum of each row, each column, and the diagonal is 34.

| | | | |
|---|---|---|---|
| 7 | 12 | 1 | 14 |
| 2 | 13 | 8 | 11 |
| 16 | 3 | 10 | 5 |
| 9 | 6 | 15 | 4 |

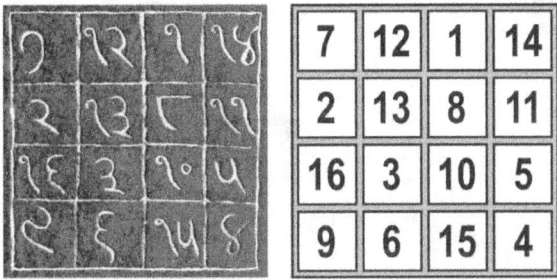

**Figure 2.39.** Chautisa Yantra

In Figures 2.40–2.42, with the two types of shading, we can see how the sum of all the rows, columns and diagonals are each 34.

| | | | |
|---|---|---|---|
| 7 | 12 | 1 | 14 |
| 2 | 13 | 8 | 11 |
| 16 | 3 | 10 | 5 |
| 9 | 6 | 15 | 4 |

**Figure 2.40**

| | | | |
|---|---|---|---|
| 7 | 12 | 1 | 14 |
| 2 | 13 | 8 | 11 |
| 16 | 3 | 10 | 5 |
| 9 | 6 | 15 | 4 |

**Figure 2.41**

| 7 | 12 | 1 | 14 |
|---|----|---|----|
| 2 | 13 | 8 | 11 |
| 16 | 3 | 10 | 5 |
| 9 | 6 | 15 | 4 |

**Figure 2.42**

What makes this particular magic square extra special is that there are other combinations of four cells that provide the sum of 34. We show these with light and dark shadings as clusters of 2 × 2 squares in Figures 2.43 and 2.44.

| 7 | 12 | 1 | 14 |
|---|----|---|----|
| 2 | 13 | 8 | 11 |
| 16 | 3 | 10 | 5 |
| 9ff | 6 | 15 | 4 |

**Figure 2.43**

| 7 | 12 | 1 | 14 |
|---|----|---|----|
| 2 | 13 | 8 | 11 |
| 16 | 3 | 10 | 5 |
| 9 | 6 | 15 | 4 |

**Figure 2.44**

In Figure 2.45, we see the sum of the opposite pairs producing the number 34.

| 7 | 12 | 1 | 14 |
|---|----|---|----|
| 2 | 13 | 8 | 11 |
| 16 | 3 | 10 | 5 |
| 9 | 6 | 15 | 4 |

**Figure 2.45**

Figures 2.46 and 2.47 show still further sums of 34.

| 7 | 12 | 1 | 14 |
|---|----|---|----|
| 2 | 13 | 8 | 11 |
| 16 | 3 | 10 | 5 |
| 9 | 6 | 15 | 4 |

| 7 | 12 | 1 | 14 |
|---|----|---|----|
| 2 | 13 | 8 | 11 |
| 16 | 3 | 10 | 5 |
| 9 | 6 | 15 | 4 |

**Figure 2.46**

| 7 | 12 | 1 | 14 |
|---|----|---|----|
| 2 | 13 | 8 | 11 |
| 16 | 3 | 10 | 5 |
| 9 | 6 | 15 | 4 |

| 7 | 12 | 1 | 14 |
|---|----|---|----|
| 2 | 13 | 8 | 11 |
| 16 | 3 | 10 | 5 |
| 9 | 6 | 15 | 4 |

**Figure 2.47**

# The Dürer Magic Square

We know that there are 880 magic squares of size 4 × 4. There is one magic square, however, that stands out from among the rest for its beauty and additional properties — not to mention its curious appearance. This particular magic square has many properties beyond those required for a square arrangement of numbers to be considered "magic." This magic square even comes to us through art, and not through the usual mathematical channels. It is depicted in the background of the famous 1514 engraving by the renowned German artist Albrecht Dürer (1471–1528), who lived in Nuremberg, Germany (see Figure 2.48).

**Figure 2.48.**   *Melencolia I*, engraving, Albrecht Dürer (1514)

As we begin to examine the magic square in Dürer's etching, we should take note that most of Dürer's works were signed by him with his initials, one over the other, and including the year in which the work was made. Here we find it in the dark-shaded region near the lower right side of the picture and highlighted in Figure 2.49, which tells us that Dürer made it in the year 1514.

**Figure 2.49.**   Initials AD of Albrecht Dürer and the year 1514

The observant reader may notice that the two center cells of the bottom row of the Dürer magic square depict the year as well. Let us examine this magic square more closely (see Figure 2.50).

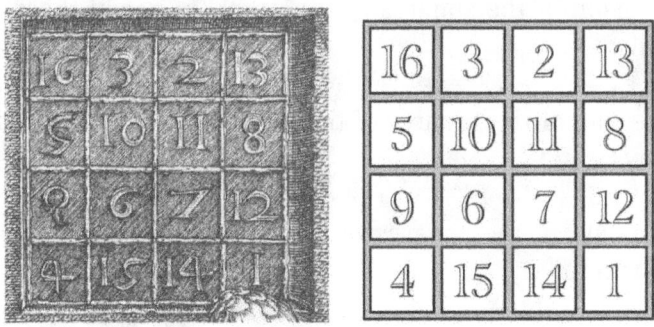

| 16 | 3 | 2 | 13 |
|----|----|----|----|
| 5 | 10 | 11 | 8 |
| 9 | 6 | 7 | 12 |
| 4 | 15 | 14 | 1 |

**Figure 2.50.**   Dürers magic square

First let's make sure that it is, in fact, a true magic square. When we calculate the sum of each of the rows, columns, and diagonals, we always get the sum of 34. That is, all that would be required for this square matrix of numbers to be considered a "magic square." However, this "Dürer Magic Square" has lots more properties that other magic squares do not have. Let us now marvel about some of these extra properties.

- The four corner numbers have a sum of 34:
  $16 + 13 + 1 + 4 = 34$
- Each of the four corner $2 \times 2$ squares has a sum of 34:
  $16 + 3 + 5 + 10 = 34$
  $2 + 13 + 11 + 8 = 34$
  $9 + 6 + 4 + 15 = 34$
  $7 + 12 + 14 + 1 = 34$
- The center $2 \times 2$ square has a sum of 34:
  $10 + 11 + 6 + 7 = 34$
- The sum of the numbers in the diagonal cells equals the sum of the numbers in the cells not in the diagonal:
  $16 + 10 + 7 + 1 + 4 + 6 + 11 + 13 = 3 + 2 + 8 + 12 + 14 + 15 + 9 + 5 = 68$
- The sum of the squares of the numbers in both diagonal cells is
  $16^2 + 10^2 + 7^2 + 1^2 + 4^2 + 6^2 + 11^2 + 13^2 = 748$
  This number is equal to
  - the sum of the squares of the numbers not in the diagonal cells:
    $3^2 + 2^2 + 8^2 + 12^2 + 14^2 + 15^2 + 9^2 + 5^2 = 748$
  - the sum of the squares of the numbers in the first and third rows:
    $16^2 + 3^2 + 2^2 + 13^2 + 9^2 + 6^2 + 7^2 + 12^2 = 748$
  - the sum of the squares of the numbers in the second and fourth rows:
    $5^2 + 10^2 + 11^2 + 8^2 + 4^2 + 15^2 + 14^2 + 1^2 = 748$
  - the sum of the squares of the numbers in the first and third columns:
    $16^2 + 5^2 + 9^2 + 4^2 + 2^2 + 11^2 + 7^2 + 14^2 = 748$
  - the sum of the squares of the numbers in the second and fourth columns:
    $3^2 + 10^2 + 6^2 + 15^2 + 13^2 + 8^2 + 12^2 + 1^2 = 748$
- The sum of the cubes of the numbers in the diagonal cells equals the sum of the cubes of the numbers not in the diagonal cells:

$16^3 + 10^3 + 7^3 + 1^3 + 4^3 + 6^3 + 11^3 + 13^3 = 3^3 + 2^3 + 8^3 + 12^3 + 14^3 + 15^3 + 9^3 + 5^3 = 9248$

- Notice the following beautiful symmetries:

$2 + 8 + 9 + 15 = 3 + 5 + 12 + 14 = 34$

$2^2 + 8^2 + 9^2 + 15^2 = 3^2 + 5^2 + 12^2 + 14^2 = 374$

$2^3 + 8^3 + 9^3 + 15^3 = 3^3 + 5^3 + 12^3 + 14^3 = 4624$

- Adding the first row to the second, and the third row to the fourth, produces a pleasing symmetry:

| $16 + 5 = 21$ | $3 + 10 = 13$ | $2 + 11 = 13$ | $13 + 8 = 21$ |
|---|---|---|---|
| $9 + 4 = 13$ | $6 + 15 = 21$ | $7 + 14 = 21$ | $12 + 1 = 13$ |

- Adding the first column to the second, and the third column to the fourth, produces a pleasing symmetry:

| $16 + 3 = 19$ | $2 + 13 = 15$ |
|---|---|
| $5 + 10 = 15$ | $11 + 8 = 19$ |
| $9 + 6 = 15$ | $7 + 12 = 19$ |
| $4 + 15 = 19$ | $14 + 1 = 15$ |

A motivated reader may wish to search for other patterns in this beautiful magic square. Remember, this is not a typical magic square, where all that is required is that all the rows, columns, and diagonals have the same sum. This Dürer magic square has many more properties. Likewise, it is worthwhile to explore the Chautisa Yantra of Figure 2.39 in order to find additional properties.

## General Properties of Magic Squares

You might wonder how it could be that both the Chautisa Yantra and the Dürer magic square have 34 as their "magic number." But, actually, this would necessarily be the case for any $4 \times 4$-magic square that uses the numbers from 1 to 16. The sum of these numbers is $1 + 2 + 3 + ... + 16 = 136$. In a magic square, every row of numbers contributes exactly a quarter of this sum. Because there are 4 rows and all rows are required to have the same sum; that sum is $136 \div 4 = 34$. By the definition of a magic square, the sum of the numbers in each row, each column, and each diagonal of the $4 \times 4$ magic square must be 34.

In that way, we can even obtain a formula for the magic number of any $n \times n$-magic square. For this, we remind you that in Chapter 1 we found that the sum of the first $n$ natural numbers is a triangle number $T_n$, and is determined by the formula:

$$T_n = 1 + 2 + 3 + \cdots + (n-1) + n = \frac{n}{2}(n+1)$$

A magic square of size $n \times n$ contains all the natural numbers from 1 to $n^2$. Applying the formula above for this situation, we find that the sum of natural numbers from 1 to $n^2$ is

$$T_{(n^2)} = \frac{n^2}{2}(n^2 + 1)$$

However, if it is required that each of the $n$ rows must have the same sum $S_n$, then the sum of each row must be $\frac{1}{n}$th of that sum, or $S_n = \frac{T_{(n^2)}}{n} = \frac{n}{2}(n^2 + 1)$.

And, in a magic square, this number must be the sum of any row, column, or diagonal.

For $n = 3$, this formula indeed gives the magic number of the earlier discussed $3 \times 3$ magic square:

$$S_3 = \frac{3}{2}(9+1) = 15$$

Here we will consider magic squares consisting of all numbers from 1 to $n^2$, where $n$, the number of row or columns, is called the *order* of the magic square. However, if one adds a constant number $k$ to all numbers in a magic square, one would obtain another magic square, with numbers ranging from $k + 1$ to $k + n^2$, and with magic number $kn + S_n$. Similarly, *multiplying* each number of a magic square with a constant $k$ would give a magic square with magic number $kS_n$.

The question that would logically be asked is how does one construct a magic square? How did Dürer come up with this special magic square? According to their order, we distinguish three types:

(a) magic squares of odd order (*n* is an odd number),

(b) magic squares of doubly-even order (*n* is a multiple of 4),

(c) magic squares of singly-even order (*n* is a multiple of 2, but not 4).

The Dürer magic square is a doubly even magic square.

## How to Construct a Doubly Even Magic Square

Since we have the Dürer square at hand, we will begin by discussing the construction of the doubly even magic squares (doubly even means that it is divisible by 4). Let us begin with the smallest of these, namely, those with four rows and four columns. We begin our construction of this doubly even magic square by first placing the numbers in the square in numerical order, as shown in the left-side square arrangement of Figure 2.51.

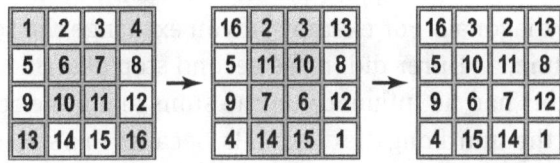

**Figure 2.51.**  Constructing Dürers square in three steps

This is not yet a magic square, because all the small numbers are in the first row, and the large numbers are in the last row. But a quick inspection shows that the sum along each diagonal has already the required value 34. Any rearrangement of the numbers within a diagonal will not change their sum. In the next step, we must try to get some of the large numbers into the upper part of the square. To do this we will exchange within the first diagonal (the "main diagonal") the numbers 1 and 16 and the numbers 6 and 11. Similarly, in the secondary diagonal, we will exchange the numbers 13 and 4 as well as the numbers 10 and 7. The cells to be changed are shaded in the left-side square of Figure 2.51. This produces the center square of

Figure 2.51. We now have been able to get some large numbers in the first row, and, indeed, the sum of along the first row is still 34! Quickly checking the remaining rows and columns reveals that this square is indeed a magic square. (No need to check the diagonals again, because exchanging numbers within a diagonal does not change its sum!) Thus, we have constructed our first magic square! However, the magic square we have obtained is *not* the same as the one that Dürer pictured in his Melencolia etching. Dürer apparently interchanged the positions of the two middle columns to allow his square to show the date that the picture was made, namely, 1514 in the middle two cells of the bottom row. This resulting arrangement of numbers is shown as the right-side square in Figure 2.51, which is Dürer's magic square, and actually has many more properties than the magic square constructed in the first step.

Once you have obtained a magic square, you can try to generate a new one by starting with the given magic square. Any change you apply to an existing magic square should not change the sums of rows, columns, and diagonals. For example, if you exchange the second and the third column, as Dürer did in the second step described above in Figure 2.51, this had no influence on the sums along the rows. But it might change the sum along the diagonals, because this step exchanges numbers in the diagonals. In general, this can be repaired by switching the second and third *row*, which does not change the sum along the columns, but restores the numbers in the diagonals. You might want to try this for the Chautisa Yantra of Figure 2.39. This one would not remain a magic square, if only the two central columns are exchanged. There you would have to exchange the central rows as well. Dürer's square is again special, in that it remains a magic square when columns 2 and 3 are exchanged (and also, if you exchange columns 1 and 4 or rows 1 and 4 or rows 2 and 3).

In general, exchanging the columns 1 and 4 (respectively, the columns 2 and 3) and then exchanging the corresponding rows would preserve the "magic property" of a square.

Another general method to create a new magic square from an existing one is to replace each number by its complement.

The complement of a number *a* in an $n \times n$ magic square is a number *b* such that $a + b = n^2 + 1$. In a square of order 4, two numbers are complementary, if their sum is 17. Therefore, looking back we can see that the first step in Figure 2.51 can also be described as the replacement of the numbers in the diagonals by their complements.

You may wish to generate new magic squares using this technique. There is a total of 880 possible magic squares of order 4. By the way, there is no magic square of order 2, and there is essentially only one magic square of order 3, the square shown in Figure 2.31, because all other magic squares of order 3 can be obtained by rotation or reflection of this original magic square.

The next larger doubly even magic square is of order 8, that is, with eight rows and columns. Once again, we place the numbers in the cells in proper numerical order as shown in Figure 2.52.

| 1 | 2 | 3 | 4 | 5 | 6 | 7 | 8 |
|---|---|---|---|---|---|---|---|
| 9 | 10 | 11 | 12 | 13 | 14 | 15 | 16 |
| 17 | 18 | 19 | 20 | 21 | 22 | 23 | 24 |
| 25 | 26 | 27 | 28 | 29 | 30 | 31 | 32 |
| 33 | 34 | 35 | 36 | 37 | 38 | 39 | 40 |
| 41 | 42 | 43 | 44 | 45 | 46 | 47 | 48 |
| 49 | 50 | 51 | 52 | 53 | 54 | 55 | 56 |
| 57 | 58 | 59 | 60 | 61 | 62 | 63 | 64 |

**Figure 2.52**

This time, we will once again replace the numbers in the diagonals with their complement — in this case, the complement of a number is the number that will produce a sum of 65. However, the diagonals in this case are the diagonals of each of the $4 \times 4$ squares included in the $8 \times 8$ square — here they are the shaded numbers. The completed magic square with all the appropriate cell changes is shown in Figure 2.53.

| 64 | 2 | 3 | 61 | 60 | 6 | 7 | 57 |
|----|----|----|----|----|----|----|----|
| 9 | 55 | 54 | 12 | 13 | 51 | 50 | 16 |
| 17 | 47 | 46 | 20 | 21 | 43 | 42 | 24 |
| 40 | 26 | 27 | 37 | 36 | 30 | 31 | 33 |
| 32 | 34 | 35 | 29 | 28 | 38 | 39 | 25 |
| 41 | 23 | 22 | 44 | 45 | 19 | 18 | 48 |
| 49 | 15 | 14 | 52 | 53 | 11 | 10 | 56 |
| 8 | 58 | 59 | 5 | 4 | 62 | 63 | 1 |

**Figure 2.53**

# Construction of a Magic Square of Order Three

We began our discussion of magic squares with the $3 \times 3$ square. As promised, we now will consider the construction of all possible $3 \times 3$ magic squares. We begin by considering the matrix of letters representing the numbers from 1 to 9 shown in Figure 2.54. Here the sums of the rows, columns, and diagonals, are denoted by $r_i$, $c_i$, and $d_j$, respectively. In a magic square of order 3, all these number sums would be equal to the magic number 15.

$$
\begin{array}{ccccc}
d_1 & c_1 & c_2 & c_3 & d_2 \\
\end{array}
$$

| | a | b | c |
|----|----|----|----|
| $r_1$ | a | b | c |
| $r_2$ | d | e | f |
| $r_3$ | g | h | i |

**Figure 2.54**

In a magic square, we would thus have $r_2 + c_2 + d_1 + d_2 = 15 + 15 + 15 + 15 = 60$.

However, this sum can also be written as

$$r_2 + c_2 + d_1 + d_2 = (d + e + f) + (b + e + h) + (a + e + i) + (c + e + g)$$
$$= 3\,e + (a + b + c + d + e + f + g + h + i) = 3\,e + 45$$

Therefore, $3e + 45 = 60$, and $e = 5$. Thus, it is established that the center position of a magic square of order 3 must be occupied by the number 5.

Recall that 2 numbers of an $n$th-order magic square are said to be complementary if their sum is $n^2 + 1$. In a $3 \times 3$-magic square, 2 numbers are complementary, if their sum is $9 + 1 = 10$. We can now see that numbers on opposite sides of 5 are complementary. For example, $a + i = d_1 - e = 15 - 5 = 10$, and, therefore, $a$ and $i$ are complementary. But so are the pairs $g$ and $c$, $b$ and $h$, and $d$ and $f$.

Let us now try to put 1 in a corner, as shown in Figure 2.55. Here $a = 1$, and therefore $i$ must be 9, so that the diagonal adds up to 15. Next we notice that 2, 3, and 4 cannot be in the same row (or column) as 1, since there is no natural number less than 9, which would be large enough to occupy the third position of such a row (or column). This would leave only the two shaded positions in Figure 2.55 to accommodate these three numbers (2, 3, and 4). Since this cannot be the case, our first attempt was a failure: the numbers 1 and 9 may occupy only the middle positions of a row (or column).

**Figure 2.55**

Therefore, we have to start with one of the 4 possible positions remaining for 1, for example, as we show in the first square of Figure 2.56. We note that the number 3 cannot be in the same row (or column) as 9, for the third number in such a row (or column) would again have to be 3, to obtain the required sum of 15. This is not possible, because a number can be used only once in the magic square. Additionally, we have seen above that 3 cannot be in the same row (or column) as 1. This leaves only the two shaded positions in Figure 2.56 for the number 3. The number opposite to three is always 7, because then $3 + 5 + 7 = 15$.

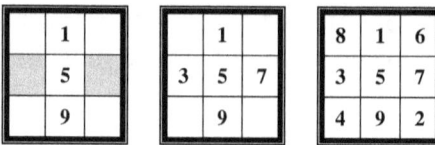

**Figure 2.56.** One of several possible magic squares

We continue with the second square in Figure 2.56, showing one of two possibilities for the placement of 3 and 7 (the other possibility has 3 and 7 exchanged). It is now easy to fill in the remaining numbers. There is only one such possibility, shown in the third square of Figure 2.56.

How many different squares are there? We could start by putting the number 1 in any of the four positions in the middle of a side. Then we have 2 possibilities for placing the 3. After that, the construction is unique. This produces the eight magic squares shown in Figure 2.57.

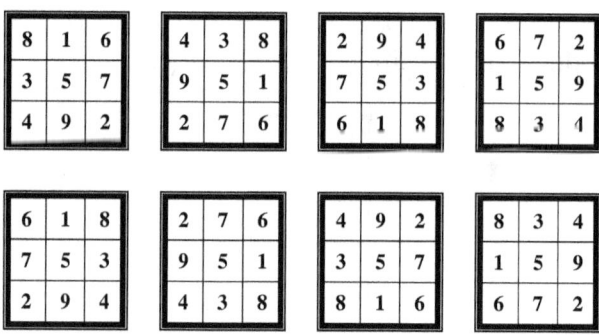

**Figure 2.57**

## Constructing Odd-Order Magic Squares

You might now want to extend this technique to construct other odd-order magic squares. The next larger magic square beyond the 3 × 3 square is that of order 5. There are many magic squares that can be made of size 5 × 5. However, this technique can become somewhat tedious. Following is a rather mechanical method for constructing an odd-order magic square.

Begin by placing a 1 in the first position of the middle column. Continue by placing the next consecutive numbers along the diagonal line in Figure 2.58.

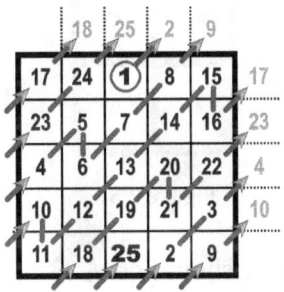

**Figure 2.58**

Whenever you drop off the square on one side, you enter again on the opposite side. So, the gray number 2 in Figure 2.58 (which now fell off the grid) must be placed in the last row. Analogously, the gray number 4 will be placed in the first column. The process continues by filling consecutively each new cell along the diagonal line until an already occupied cell is reached (as is the case with the number 6). Rather than placing a second number in an already-occupied cell, the number is placed below the previous number. The process continues until the last number is reached. After some practice with this procedure you will begin to recognize certain patterns (e.g., the last number always occupies the middle position of the bottom row). This is just one of many ways of constructing odd-order magic squares. Not counting rotations and reflections, there are 275,305,224 different 5 × 5-magic squares. Exact numbers for higher order magic squares are unknown.

## Creating Singly-Even Order Magic Squares

A different technique is used to construct magic squares of singly even order (i.e., where the number of rows and columns is even, but not a multiple of 4). Any singly even order (say, of order $n$) magic square may be separated into quadrants (Figure 2.59). For convenience, we will label these quadrants as A, B, C, and D.

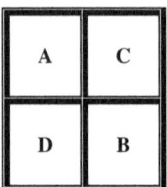

**Figure 2.59**

The order of the square should be $n$, a singly-even number, hence the order of each of the quadrants must be odd. We denote the order of the quadrant by $k = 2m + 1$ (which is always an odd number for $m = 1, 2, 3, ...$). As there is no magic square of order 2, the smallest singly-even-order magic square will have order 6, in which case $m = 1$ and $k = 3$.

We have $n = 2k = 2(2m + 1) = 6, 10, 14, 18, ...$ (for $m = 1, 2, 3, 4, ...$)

Each of the four quadrants contains $k^2$ different numbers. We start by creating a magic square of odd-order $k$ according to the method described earlier. For $n = 6$ and $k = 3$, the starting point will, therefore, be one of the variants of the $3 \times 3$ magic square. We will choose the first of the magic squares shown in Figure 2.57.

We begin by entering this magic square into quadrant A. The magic squares in quadrants B, C, and D will be obtained as shown for the case $n = 6$ in Figure 2.60.

| 8 | 1 | 6 | 18+8 | 18+1 | 18+6 |
|---|---|---|------|------|------|
| 3 | 5 | 7 | 18+3 | 18+5 | 18+7 |
| 4 | 9 | 2 | 18+4 | 18+9 | 18+2 |
| 27+8 | 27+1 | 27+6 | 9+8 | 9+1 | 9+6 |
| 27+3 | 27+5 | 27+7 | 9+3 | 9+5 | 9+7 |
| 27+4 | 27+9 | 27+2 | 9+4 | 9+9 | 9+2 |

**Figure 2.60**

Here square B is obtained by adding $k^2$ to all numbers of square A. Square C is obtained by adding $k^2$ to all numbers of square B, and square D is obtained by adding $k^2$ to all numbers of square C.

Recall that adding a fixed number to all numbers of a magic square does not change the magic property: The sum of rows, columns, and diagonals would still remain the same. Thus, the squares B, C, and D are also magic squares, only they do not use the numbers from 1 to $k^2$. For example, square B uses instead the numbers from $k^2 + 1$ to $2k^2$ (for $n = 6$, the numbers from 10 to 18). The square of order $n$ obtained in this way is shown in Figure 2.61. Although it has magic squares in its quadrants, it is not yet a magic square itself.

| 8 | 1 | 6 | 26 | 19 | 24 |
|---|---|---|----|----|----|
| 3 | 5 | 7 | 21 | 23 | 25 |
| 4 | 9 | 2 | 22 | 27 | 20 |
| 35 | 28 | 33 | 17 | 10 | 15 |
| 30 | 32 | 34 | 12 | 14 | 16 |
| 31 | 36 | 29 | 13 | 18 | 11 |

**Figure 2.61**

Continuing along with our construction of the singly even order magic square, we now have to make some adjustments to the square we have developed to this point. Recall that the integer $m$ determines the order through the formula $n = 2\,(2m + 1)$.

In general, the adjustments will be the following: We first take the numbers in the first $m$ positions in each row of quadrant A, except the middle row, where we will skip the first position and then take the next $m$ positions. Then we will exchange the numbers in these positions with the correspondingly placed numbers in square D, below. We then take the last $m - 1$ cells in each row of square C and then exchange them with the numbers in the corresponding cells of square B. For $n = 6$ and $m = 1$, the positions in squares A and D that will be changed during that procedure are shaded in Figure 2.61.

Since, in this case, $m - 1 = 0$, the squares B and C on the right side remain unaltered. The resulting square is shown in Figure 2.62. You may verify that it is indeed a magic square.

| 35 | 1 | 6 | 26 | 19 | 24 |
|----|----|----|----|----|----|
| 3 | 32 | 7 | 21 | 23 | 25 |
| 31 | 9 | 2 | 22 | 27 | 20 |
| 8 | 28 | 33 | 17 | 10 | 15 |
| 30 | 5 | 34 | 12 | 14 | 16 |
| 4 | 36 | 29 | 13 | 18 | 11 |

**Figure 2.62**

We illustrate this procedure once again with the next larger singly even magic square, which is of order $n = 10$ and in this case, $m = 2$.

(1) Starting with a magic square of order 5, we take the one created by the method explained previously (Figure 2.58).

(2) We fill the four quadrants of the $n \times n$ square. We create square B by adding 25 to all numbers of square A, and then continue as we did earlier. The result is shown as the first square in Figure 2.63.

(3) Take the first 2 positions of each row of quadrant A, except the middle row, where you skip the first cell and then take the next 2 positions. Exchange the numbers in these cells with the numbers in the corresponding cells of square D. Figure 2.63 has the corresponding positions shaded.

| 17 | 24 | 1 | 8 | 15 | 67 | 74 | 51 | 58 | 65 |
|----|----|----|----|----|----|----|----|----|----|
| 23 | 5 | 7 | 14 | 16 | 73 | 55 | 57 | 64 | 66 |
| 4 | 6 | 13 | 20 | 22 | 54 | 56 | 63 | 70 | 72 |
| 10 | 12 | 19 | 21 | 3 | 60 | 62 | 69 | 71 | 53 |
| 11 | 18 | 25 | 2 | 9 | 61 | 68 | 75 | 52 | 59 |
| 92 | 99 | 76 | 83 | 90 | 42 | 49 | 26 | 33 | 40 |
| 98 | 80 | 82 | 89 | 91 | 48 | 30 | 32 | 39 | 41 |
| 79 | 81 | 88 | 95 | 97 | 29 | 31 | 38 | 45 | 47 |
| 85 | 87 | 94 | 96 | 78 | 35 | 37 | 44 | 46 | 28 |
| 86 | 93 | 100 | 77 | 84 | 36 | 43 | 50 | 27 | 34 |

| 92 | 99 | 1 | 8 | 15 | 67 | 74 | 51 | 58 | 40 |
|----|----|----|----|----|----|----|----|----|----|
| 98 | 80 | 7 | 14 | 16 | 73 | 55 | 57 | 64 | 41 |
| 4 | 81 | 88 | 20 | 22 | 54 | 56 | 63 | 70 | 47 |
| 85 | 87 | 19 | 21 | 3 | 60 | 62 | 69 | 71 | 28 |
| 86 | 93 | 25 | 2 | 9 | 61 | 68 | 75 | 52 | 34 |
| 17 | 24 | 76 | 83 | 90 | 42 | 49 | 26 | 33 | 65 |
| 23 | 5 | 82 | 89 | 91 | 48 | 30 | 32 | 39 | 66 |
| 79 | 6 | 13 | 95 | 97 | 29 | 31 | 38 | 45 | 72 |
| 10 | 12 | 94 | 96 | 78 | 35 | 37 | 44 | 46 | 53 |
| 11 | 18 | 100 | 77 | 84 | 36 | 43 | 50 | 27 | 59 |

**Figure 2.63**

(4) To complete the magic square, we take last $m - 1$ positions (here the last positions, since $m - 1 = 1$) in each row of the squares C and B and interchange them. This gives us the magic square shown as the second square in Figure 2.63.

We now have a procedure for constructing each of the three types of magic squares: the odd-order magic square, and both the singly-even-order magic squares, and the doubly-even-order magic squares.

## An Alphamagic Square

We end this discussion about magic squares with a curiosity, just for entertainment. You can verify that the first square in Figure 2.64 is a magic square. The sum of its rows, columns, and diagonals is 45.

| 12 | 28 | 5 |
|----|----|----|
| 8 | 15 | 22 |
| 25 | 2 | 18 |

| twelve | twenty eight | five |
|--------|--------------|------|
| eight | fifteen | twenty two |
| twenty five | two | eighteen |

| 6 | 11 | 4 |
|----|----|----|
| 5 | 7 | 9 |
| 10 | 3 | 8 |

**Figure 2.64.** An alphamagic square

However, it has an additional property that makes it a so-called *alphamagic* square. Replace the numbers by their written words. The number of letters in each word generates a new magic square — the third square in Figure 2.64. You can convince yourself of its magic property either by computing all sums of the rows, columns, and diagonals, or by noticing that it can also be obtained from the 3 × 3 magic square by adding 2 to all its numbers. (Remember, adding a constant number to all numbers of a magic square generates a new magic square.)

## A Sequential Square

After having gone through this extensive presentation of magic squares, where all the rows and all the columns and diagonals have the same sum, it would be a good challenge for the audience to seek out a 4×4 square arrangement of numbers that were the sums of the rows and columns and diagonals are all different, and in particular each one comes up with a number, which would form a set of sequential numbers. One solution is offered in Figure 2.65, where the sums of the rows, columns and diagonals are shown to form the sequence: 30, 31, 32, 33, 34, 35, 36, 37, 38, 39.

| | 15 | 2 | 12 | 4 | 33 |
|---|---|---|---|---|---|
| | 1 | 14 | 10 | 5 | 30 |
| | 8 | 9 | 3 | 16 | 36 |
| | 11 | 13 | 6 | 7 | 37 |
| 34 | 35 | 38 | 31 | 32 | 39 |

**Figure 2.65**

You now have provided ample entertainment for your audience with regard to magic squares of odd-order, singly-even-order, doubly-even-order, and then non-magic squares of curious types. Perhaps this will lead the audience to seek further investigation into square arrangement of numbers such as the following.

## A Magic Square of Squares

Oftentimes, famous mathematicians such as the Swiss mathematician Leonhard Euler (1707–1783) provide us with entertaining aspects of mathematics. In 1770, Euler produced a magic square where all the entries are square numbers. In Figure 2.66, Euler's magic square has a sum of 8515 in all the rows, columns, and diagonals. This square intentionally does not use consecutive numbers.

| | | | |
|---|---|---|---|
| $68^2$ | 292 | $41^2$ | $37^2$ |
| $17^2$ | $31^2$ | $79^2$ | $32^2$ |
| $59^2$ | $28^2$ | $23^2$ | $61^2$ |
| $11^2$ | $77^2$ | $8^2$ | $49^2$ |

**Figure 2.66**

To date, it is not known whether a $3 \times 3$ magic square could be made consisting of all square numbers. However, the French mathematician Christian Boyer has created magic squares consisting of square numbers which are $5 \times 5$ magic squares, $6 \times 6$ magic squares, and a special magic square which is a $7 \times 7$ magic square where the numbers are all consecutive integers from $0^2 - 48^2$ as shown in Figure 2.67.

| | | | | | | |
|---|---|---|---|---|---|---|
| $25^2$ | $45^2$ | $15^2$ | $14^2$ | $44^2$ | $5^2$ | $20^2$ |
| $16^2$ | $10^2$ | $22^2$ | $6^2$ | $46^2$ | $26^2$ | $42^2$ |
| $48^2$ | $9^2$ | $18^2$ | $41^2$ | $27^2$ | $13^2$ | $12^2$ |
| $34^2$ | $37^2$ | $31^2$ | $33^2$ | $0^2$ | $29^2$ | $4^2$ |
| $19^2$ | $7^2$ | $36^2$ | $302$ | $1^2$ | $36^2$ | $40^2$ |
| $21^2$ | $32^2$ | $2^2$ | $39^2$ | $23^2$ | $43^2$ | $8^2$ |
| $17^2$ | $28^2$ | $47^2$ | $3^2$ | $11^2$ | $24^2$ | $38^2$ |

**Figure 2.67**

For more entertainment, we can take this a step further to find a magic square that is as shown in Figure 2.68 and also when each of the numbers in the cells are squared then a magic square is still retained. One such was produced by the Polish mathematician Jaroslaw Wroblewski, where the sum of the rows, columns, and diagonals of the initial magic square is 408, which is shown in Figure 2.68. For the magic square that contains the squares of these numbers, the sum of the rows, columns, and diagonals is 36,826. This is shown in Figure 2.69.

| | | | | | |
|---|---|---|---|---|---|
| 17 | 36 | 55 | 124 | 62 | 114 |
| 58 | 40 | 129 | 50 | 111 | 20 |
| 108 | 135 | 34 | 44 | 38 | 49 |
| 87 | 98 | 92 | 102 | 1 | 28 |
| 116 | 25 | 86 | 7 | 96 | 78 |
| 22 | 74 | 12 | 81 | 100 | 119 |

**Figure 2.68**

| $17^2$ | $36^2$ | $55^2$ | $124^2$ | $62^2$ | $114^2$ |
|--------|--------|--------|---------|--------|---------|
| $58^2$ | $40$ | $129^2$ | $50^2$ | $111^2$ | $20^2$ |
| $108^2$ | $135^2$ | $34^2$ | $44^2$ | $38^2$ | $49^2$ |
| $87^2$ | $98^2$ | $92^2$ | $102^2$ | $1^2$ | $28^2$ |
| $116^2$ | $25^2$ | $86^2$ | $7^2$ | $96^2$ | $78^2$ |
| $22^2$ | $74^2$ | $12^2$ | $81^2$ | $100^2$ | $119^2$ |

**Figure 2.69**

If you really want to impress your audience, you can take this one step further to show them that there are magic squares consisting of cubes, such as one created by Lee Morgenstern, which has only the rows and columns with the same sum as shown in Figure 2.70. The sum of each row and column is 7,095,816. An ambitious audience may want to verify that!

| $16^3$ | $20^3$ | $18^3$ | $192^3$ |
|--------|--------|--------|---------|
| $180^3$ | $81^3$ | $90^3$ | $15^3$ |
| $108^3$ | $135^3$ | $150^3$ | $9^3$ |
| $2^3$ | $160^3$ | $144^3$ | $24^3$ |

**Figure 2.70**

Actually, the smallest magic square consisting of cubes is an $8 \times 8$ magic square that was developed in 2008 by German mathematics teacher Walter Trump, which we show in Figure 2.71. The sum of all the rows, columns, and diagonals is six hundredths 636,363.

| $11^3$ | $9^3$ | $15^3$ | $61^3$ | $18^3$ | $40^3$ | $27^3$ | $68^3$ |
|------|------|------|------|------|------|------|------|
| $21^3$ | $34^3$ | $64^3$ | $57^3$ | $32^3$ | $24^3$ | $45^3$ | $14^3$ |
| $38^3$ | $3^3$ | $58^3$ | $8^3$ | $66^3$ | $2^3$ | $46^3$ | $10^3$ |
| $63^3$ | $31^3$ | $41^3$ | $30^3$ | $13^3$ | $42^3$ | $39^3$ | $50^3$ |
| $37^3$ | $51^3$ | $12^3$ | $6^3$ | $34^3$ | $65^3$ | $23^3$ | $19^3$ |
| $47^3$ | $36^3$ | $43^3$ | $33^3$ | $29^3$ | $59^3$ | $52^3$ | $4^3$ |
| $55^3$ | $53^3$ | $20^3$ | $49^3$ | $25^3$ | $16^3$ | $5^3$ | $56^3$ |
| $1^3$ | $62^3$ | $26^3$ | $35^3$ | $48^3$ | $7^3$ | $60^3$ | $22^3$ |

**Figure 2.71**

Although this may be a bit beyond what can be expected of most general audiences, it still could be considered entertaining and perhaps among some generate further investigation to other analogous magic squares.

## Magic Multiplication Square

Perhaps it is more difficult, but nevertheless challenging in an entertaining fashion to develop a magic square were rather than to seek a common sum for each row and column, we would search for an arrangement that would yield the same product for every row and column in a $3 \times 3$ magic square. We show this in Figure 2.72. (Note: we are not including the diagonals in this magic square.)

| 1 | 12 | 10 |
|----|----|----|
| 15 | 2 | 4 |
| 8 | 5 | 3 |

**Figure 2.72**

# An Upside-Down Magic Square

To offer further entertainment, we show here a most unusual magic square. This one can be read in various positions even though the numbers will be different. That is, if you turn the magic square shown in Figure 2.73 upside down, you will still have a sum of 264 for each row, column, and diagonal.

| 96 | 11 | 89 | 68 |
|----|----|----|----|
| 88 | 69 | 91 | 16 |
| 61 | 86 | 18 | 99 |
| 19 | 98 | 66 | 81 |

**Figure 2.73**

Furthermore, there are many other combinations of cells that lead to the same sum of 264, such as the four corner cells, or the two center cells in each of the end columns, 88 + 61 + 16 + 99 = 264. There are many more combinations of four cells that lead to the sum of 264 and for entertainment your audience might like to search for them and see how many different ones can be found.

Another such magic square that can be turned upside down is shown in Figure 2.74, however this time the numbers are a bit larger and the common sum is 19,998.

| 1118 | 8181 | 1888 | 8811 |
|------|------|------|------|
| 8888 | 1811 | 8118 | 1181 |
| 8111 | 1188 | 8881 | 1818 |
| 1881 | 8818 | 1111 | 8188 |

**Figure 2.74**

Once again, there are many other combinations of four that will yield the common sum of 19,998. Just a bit more entertainment!

## A Magic Triangle

Now that we have had a rather complete set of entertainments with magic squares, we can change the agenda a bit by considering a somewhat magic triangle. Here we will seek to find numbers in the triangular arrangement shown in Figure 2.75 where the difference of every adjacent pair of numbers is represented by the number above and between. For example, in the triangular arrangement shown in Figure 2.75, we provide one example of how a pair of adjacent numbers has its difference placed above and between them.

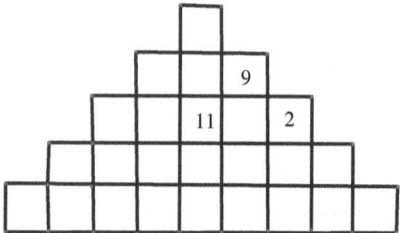

**Figure 2.75**

After your audience has had time to try to solve this problem, you might want to provide further clues, by appropriately inserting the numbers 14 and 15 which should allow them a path to present a successful solution as shown in Figure 2.76.

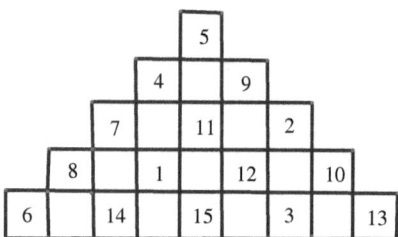

**Figure 2.76**

## A Magic Pentagram

We have now had lots of entertainment with magic squares and a magic triangle, and now we will have a magic pentagram. In Figure 2.77, we have a pentagram with all the intersection points marked. What we seek here is to replace each of the letters with numbers so that the 4 intersection points along any one of the 5 lines have a sum of 24.

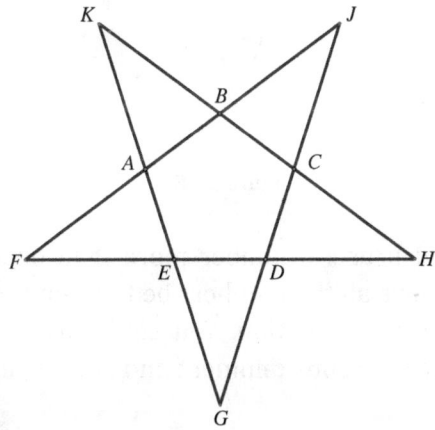

**Figure 2.77**

We should say at the outset that this problem cannot be done with any sequence of consecutive numbers. We chose the number 24, since that is the smallest number for a sum that can be used without introducing negative numbers. If you choose to entertain your audience with other challenges of this sort, you might use even numbers. In Figure 2.78, we offer a solution where the sum of every line of 4 numbers has a sum of 24.

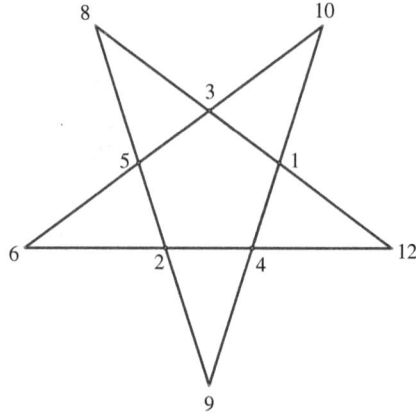

**Figure 2.78**

Although you can use a number more than once, it would, of course, be silly to have all the numbers be the same, which obviously would lead to a correct solution. You could also include negative numbers to enhance the entertainment and challenge.

## Instant Calculation

We can sometimes entertain others by showing them a little mathematical trick. The one we are about to expose is your ability to add much more quickly than the audience can possibly do. We will use this matrix of numbers shown in Figure 2.79, which has 5 columns and 6 rows. Your audience is to select one number in each of the 5 columns from this matrix of numbers and have the task of adding them to get a sum. However, you will be able to get the sum instantly following the procedure were about to expose. In short, you will add up the units digits of each of the numbers that were selected, then subtract that number from 50 and then create a 4-digit number comprised of the original sum tagged onto the difference from 50. Perhaps it is best we use an example to show how this works.

| 366 | 642 | 582 | 278 | 558 |
| --- | --- | --- | --- | --- |
| 762 | 147 | 285 | 377 | 954 |
| 69 | 345 | 186 | 872 | 756 |
| 564 | 48 | 384 | 674 | 459 |
| 168 | 246 | 87 | 575 | 657 |
| 663 | 543 | 483 | 179 | 855 |

**Figure 2.79**

Following the directions offered above, suppose your audience selects the numbers 762, 246, 87, 872, and 459, and then must add these to get a sum. In the meantime, all you need to do is to take the sum of the units digits of these numbers: $2 + 6 + 7 + 2 + 9 = 26$, and subtract that number from 50, to get $50 - 26 = 24$. Then all you need to do to get the sum of the originally selected numbers is to tag these last 2 obtained numbers (24 and 26) together as 2426, and the answer is arrived at. To ensure a good understanding, we offer another example, where we would have the audience add up the following numbers selected, one from each column: $663 + 48 + 582 + 377 + 954$. Here again, you would add the units digit of each of the number selected: $3 + 8 + 2 + 7 + 4 = 24$. Then subtract this from 50 to get 26, and then combine these 2 numbers to obtain the required some, namely, 2624. One shouldn't need too much practice to replicate this procedure.

At this point your audience may wonder how you came about doing this addition so rapidly. The trick lies in the fact that the numbers in this matrix were selected in the following way: In each column the tens digit is the same for all numbers, and the units digit plus the hundreds digit also yields the same sum for each column. (We use a zero when there is no hundreds digit.) A little reflection will expose this scheme used here and allow your audience to create similar addition tricks.

## Honoring the Opening of the United Nations Headquarters in New York City–1952

Here we offer an entertaining arrangement of numbers, where following a specific rule will always yield a sum of 1952. Consider the square arrangement of numbers shown in Figure 2.80.

| 212 | 316 | 413 | 515 | 614 |
|-----|-----|-----|-----|-----|
| 203 | 307 | 404 | 506 | 605 |
| 190 | 294 | 391 | 493 | 592 |
| 176 | 280 | 377 | 479 | 578 |
| 161 | 265 | 362 | 464 | 563 |

**Figure 2.80**

Following this technique your audience will always get a sum of 1952. This procedure is as follows: They are to select any number from the matrix shown in Figure 2.80, and then select the second number that is not in the same row or column as the previous number. They then continue to select a third number that is not in the row or column as the previous 2 numbers which were selected. Continuing along picking a fourth number which is not in the same row or column as any of the previously selected numbers, that leaves one number to select, which is not in any of the previously used rows or columns. Let's consider one possible process. Supposing we select the number 203, which is in the first column second row, which tells us that these are 2 rows and columns that cannot be used in the future. We will then pick the number 294, which no longer allows us to use the second column and the third row. Our next choice will be the number 377, once again that eliminates the third column and the fourth row for future selections. Our fourth choice will be the number 464, which no longer allows us to use the fourth column and the fifth row. The only number remaining for us to select is the number 614 which is in the first row and fifth column, which have not been used previously. We now find the sum of these numbers: $203 + 294 + 377 + 464 + 614 = 1952$. And so, we have a technique for honoring the year that the United Nations headquarters in New York was opened.

## The Logic of Weighing Coins

Here is one problem that uses logic in an entertaining fashion.

*Suppose you have 8 one-dollar coins and you know that one of the coins is counterfeit and weighs a bit more than the other 7 coins. Using a simple balance scale, how can you determine which is the counterfeit coin with exactly 2 weighings?*

Begin by breaking up the 8 coins into 3 groups, consisting of 2 groups of 3 coins each and 1 group of 2 coins. The first weighing would be by placing the 2 groups of 3 coins, each on either end of the balance scale. If they balance, then you know that the heavier coin must be in the group of 2 coins and a second weighing of these 2 coins will determine which is the heavier coin. On the other hand, if the 2 groups of 3 coins, do not balance, then take the heavier 3 coins group, and take any 2 coins and weigh them against each other. If they balance, then you know the heavier coin was the third coin of the group. If they do not balance, then you know which is the heavier coin immediately.

## Weighing Solution

Another weighing problem could be placed with a vegetable market that wishes to measure the weights from 1 pound to 40 pounds using only 4 types of weights. What would these weights have to be in order to be able to weigh each pound in the sequence by either adding or subtracting several of these 4 weights?

It turns out that with 4 types of weights of the following size, 1, 3, 9, and 27, all the pounds from 1 to 40 can be weighed. For example, to weigh a 7-pound item the weights would be $1 + 3 + 3 = 7$. To weigh a 39-pound item, it would require the following weights: $3 + 9 + 27 = 39$. Your audience should have fun trying to figure out how all of these weighings can be done with these 4 types of weights.

## The Escalator Race

Here is a tricky question that requires little bit of logical thinking and a little bit of algebra.

Two men are racing up a long escalator. One of the men runs three times faster than the other one. While running up the escalator, the faster runner counts 75 steps, while the slower runner counts 50 steps. If the escalator were to stop, how many steps would be countable (visible)?

As one is taught in school, we usually begin by letting $x$ equal that quantity that we are looking for, which in this case, the number of visible steps on the escalator. Therefore, the number of steps is the sum of the steps counted by each of the men plus those that moved away while they were running up the stairs. The rate of speed of each of the men, $r$ and $3r$, and the speed of the escalator is $R$. We can then set up the following relationships: $\frac{r}{R} = \frac{50}{x-50}$ and $\frac{3r}{R} = \frac{75}{x-75}$. One way to solve the equation would be to multiply the first equation by 3, to get $\frac{3r}{R} = \frac{150}{x-50}$, which then enables us to get the following equation: $\frac{75}{x-75} = \frac{150}{x-50}$, which then can be written as: $75(x-50) = 150(x-75)$, and then we have $x = 100$. Incidentally, the slower runner's speed is the same as the speed of the escalator.

## The Monty Hall Problem

"Let's Make a Deal" was a long-running television game show that featured a problematic situation for a randomly selected audience member to come on stage and presented with three doors to select one, hopefully the one with the car and not one of the other two doors, each of which had a donkey behind it. There was only one wrinkle in this: after the contestant made her selection, the host, Monty Hall, exposed one of the two donkeys behind a not-selected door (leaving two doors still unopened) and the audience participant was asked if she wanted to stay with her original selection (not yet revealed) or switch to the other unopened door. At this point, to heighten the suspense, the rest of the audience would shout out "stay" or "switch" with seemingly equal frequency. The question is what to do? Does it make a difference? If so, which is the better strategy (i.e., the greater probability of winning) to use here? Let us look at this now step-by-step. The result gradually will become clear.

There are *two donkeys* and *one car* behind these doors.

You must try to get the car. You select Door #3.

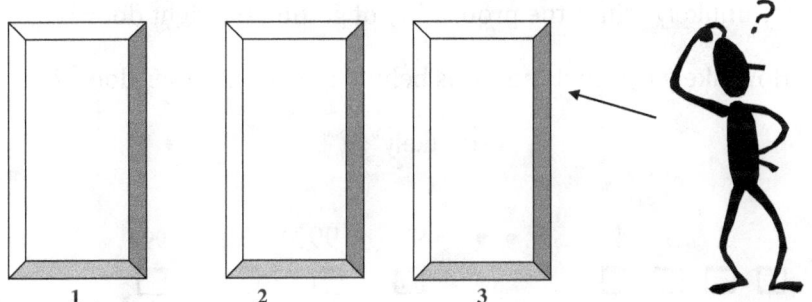

Monty Hall opens one of the doors that you did not select and exposes a donkey.

*He asks:* "Do you still want your first-choice door, or do you want to switch to the other closed door"?

To help make a decision, consider an *extreme case*:

Suppose there were 1000 doors instead of just three doors.

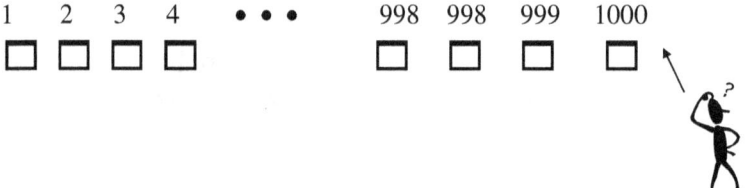

You choose Door # 1000. How likely is it that you chose the right door?

"Very unlikely" since the probability of getting the right door is $\frac{1}{1000}$

How likely is it that the car is behind one of the other doors?

"Very likely": $\frac{999}{1000}$

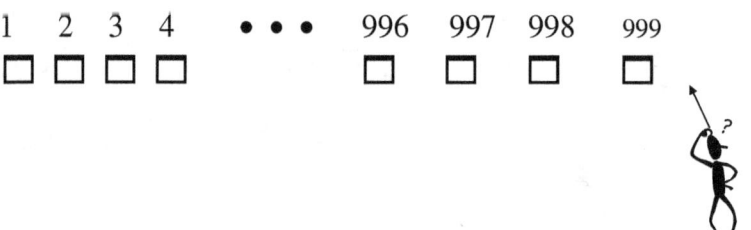

These are all "very likely" doors!

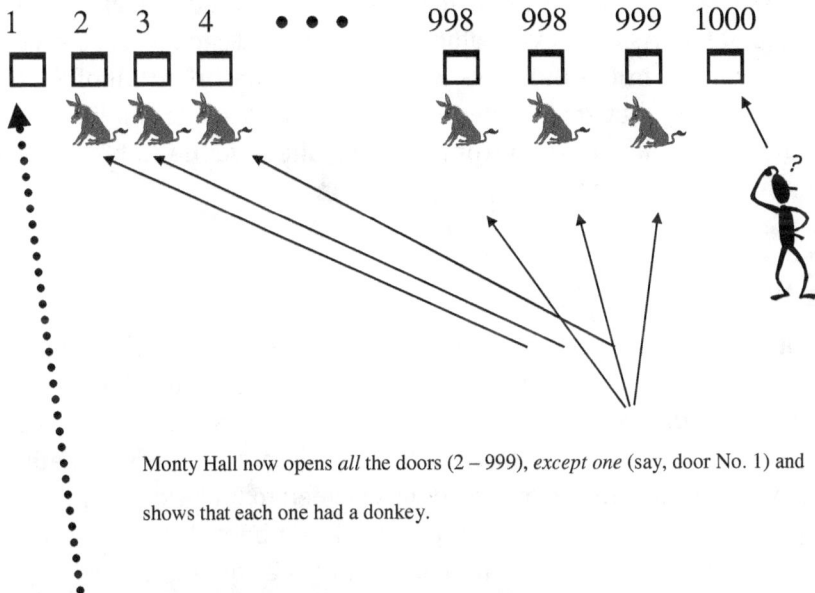

Monty Hall now opens *all* the doors (2 – 999), *except one* (say, door No. 1) and shows that each one had a donkey.

A "very likely" door is left:   Door #1.

We are now ready to answer the question. Which is a better choice:

- Door No. 1000 ("Very unlikely" door), or
- Door No. 1 ("Very likely" door)?

The answer is now obvious. We ought to select the "very likely" door, which means "switching" is the better strategy for the audience participant to follow. In the extreme case, it is much easier to see the best strategy, than had we tried to analyze the situation with the three doors. The principle is the same in either situation.

This problem has caused an argument in academic circles and was also a topic of discussion in the New York Times, and other popular publications as well. John Tierney wrote in *The New York Times* (Sunday, July 21, 1991) that "perhaps it was only an illusion, but for a moment here it seemed that an end might be in sight to the debate raging among mathematicians," readers of *Parade* magazine and fans of the television game show *Let's Make a Deal.* They began arguing after Marilyn vos Savant published a puzzle in Parade magazine. As readers of her "Ask Marilyn" column are reminded each week, Ms. vos Savant is listed in the Guinness Book of World Records Hall of Fame for Highest I.Q., but that credential did not impress the public when she answered this question from a reader. She gave the right answer, but still many mathematicians argued.

## A Tough Challenge!

It can sometimes be entertaining to challenge friends with a number puzzle, one of which we present here. In Figure 2.81, we show a circle with 10 points around it, where 4 of them are occupied by a number and 6 of them require the placement of the correct numbers. To determine what these 6 numbers are to be, we need to understand the rule for the placement of numbers. From the 4 numbers already inserted, the following condition is true. The sum of the squares of two adjacent numbers must be equal to the sum of the squares of two diametrically opposite adjacent numbers. For example, the sum of the squares of 16 and 8 is $256 + 4 = 260$, and the sum of the squares of 14 and 8 is $196 + 64 = 260$. The task ahead now is to find the appropriate numbers for the remaining open positions around the circle. The numbers we seek must be integers in one- or two-digit size.

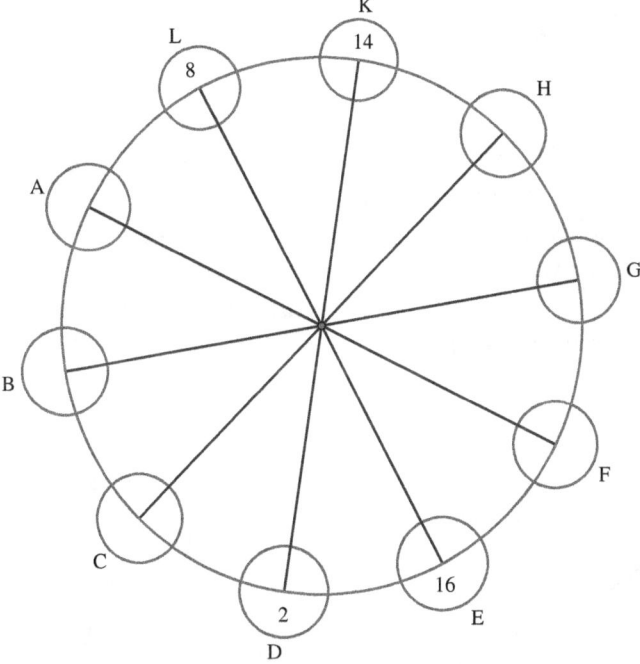

**Figure 2.81**

At first inspection, we will notice that the difference of the squares of the opposite lying numbers is a constant. That is, the difference of the squares between 16 and 8 is 256 − 64 =192, while the difference between the squares of 14 and 2 is 196 − 4 = 192, as well.

With simple algebra, we can show that the difference of the squares of 2 consecutive numbers is equal to twice the smaller number increased by 1. $x^2 - (x+1)^2 = (x-(x+1))(x+(x+1)) = 2x+1$. For example, squares of 2 consecutive numbers such as 7 and 8 have a difference of 64 − 49 = 15, which is equal to $2 \times 7 + 1 = 15$.

We also know that the difference of any 2 squares is equal to the product of the difference of the 2 numbers and the sum of the 2 numbers, as seen algebraically as $x^2 - y^2 = (x-y)(x+y)$. We see this for the difference between squares of 6 and 9 is 81 − 36 = 45, which is then also equal to $(6+9) \times (6-9) = 15 \times 3 = 45$.

Using this information, let's revert back to the number we found above, namely, 192. This number can be separated and expressed as the product of 5 different pairs of even numbers: $12 \times 16$, $8 \times 24$, $6 \times 32$, $4 \times 48$, $2 \times 96$. Since these are all even numbers, we can make them more manageable by dividing them all by 2, to get $6 \times 8$, $4 \times 12$, $3 \times 16$, $2 \times 24$, $1 \times 48$. If we now take the difference and the sum of each pair, we get the following: (14, 2), (16, 8), (19, 13), (22, 26), and (49, 47). These are then the required numbers that we need to fill our vacant spots. Four of these already exist on the circle. Therefore, we need to place the remaining 6 numbers as we show in Figure 2.82.

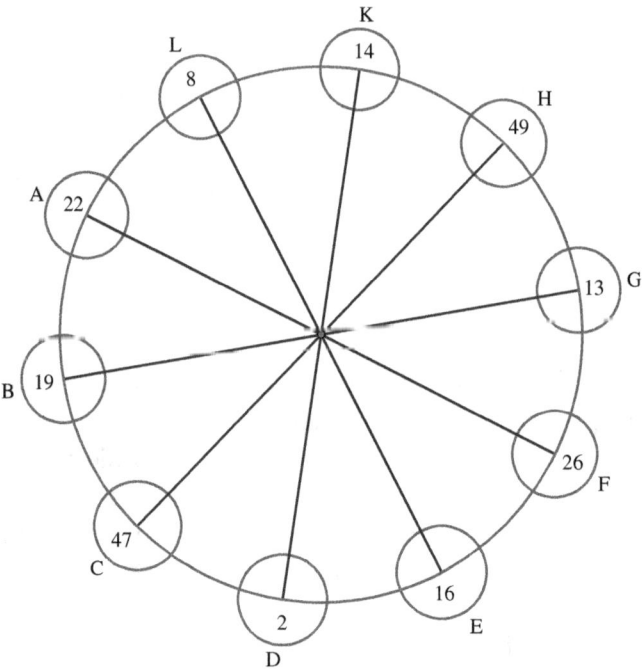

**Figure 2.82**

An ambitious reader may wish to try to set up a similar arrangement of numbers around the circle. In any case, the challenge should be entertaining and also providing an insight of number relationships, which we found by using simple elementary algebra.

Throughout this chapter, your audience has been entertained by various forms of logic as they manifest themselves in mathematics. It is quite likely that your audience will be taking some of the ideas encountered in this chapter to impress other friends, which will serve as a fine multiplier effect that will make mathematics more popular to the general audience not only as a source of entertainment, but also as a valuable aspect for logical thinking in our modern society.

## Endnote

[1] The quadratic formula for the equation $ax^2 + bx + c = 0$ is as follows $x = \frac{-b \pm \sqrt{b^2 - 4ac}}{2a}$.

# Chapter 3

# Geometric Surprises

Entertaining an audience with geometry takes on a different perspective. First of all, the fact that it's visual requires that you have a way to exhibit the various unusual aspect of geometry that you are about to share. Furthermore, it is important that the presentations you make are done in a somewhat dramatic fashion that will eventually exhibit the amazing aspects that you would like your audience to properly appreciate. We will begin with an everyday observation which very few people take note of, and yet are awed by the phenomenon. Begin with the following question: Have you ever wondered why sewer covers are always round? Well, as an entertainer you might want to let your audience ponder this question a bit. All kinds of unexpected responses can be awaited. As you can see in Figure 3.1, the answer is rather simple: the circular cover cannot fall into the hole.

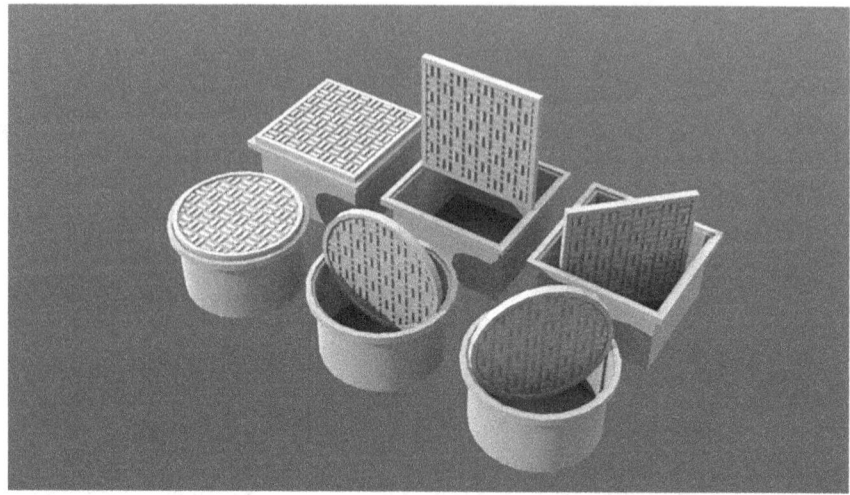

**Figure 3.1**

You can further entertain your audience by telling them that there are other shapes that also can be used as a sewer cover, and which will not fall into the hole. One of these is known as the Reuleaux triangle, named after the German engineer Franz Reuleaux (1829–1905) and which is shown in Figure 3.2. It is formed by constructing an equilateral triangle and drawing a circular arc on each side with the center of the arc's circle at the opposite vertex.

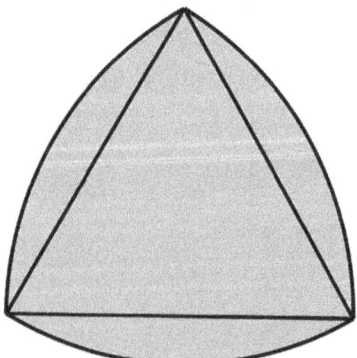

**Figure 3.2**

A motivated audience will probably seek other shapes of sewer covers that will also not fall into the hole that they are covering. One such example would be to replace the equilateral triangle with a regular pentagon and draw the circular arcs as before.

## Surrounding Pennies

By the way, speaking of circles, suppose you would like to place a certain number of pennies around a given penny so that they are all tangent to the given penny. How many pennies could be accommodated around a given penny? Unexpectedly, the answer is six pennies as you can see in Figure 3.3, where the diagram speaks for itself and then in Figure 3.4, we can see the actual result.

**Figure 3.3**

**Figure 3.4**

Now let us continue our journey through further entertaining geometry situations. Most of them should enchant your audience.

## The Golden Section Constructed by Paper-Folding

There are many things in mathematics that are "beautiful," yet, sometimes the beauty is not apparent at first sight. This is not the case with the Golden Section, which ought to be beautiful at first sight regardless of the form in which it is presented. The Golden Section refers to the proportion into which a line segment is divided by a point. Simply, for the segment $AB$, the point $P$ partitions (or divides) it into two segments, $AP$ and $PB$, such that $\frac{AP}{PB} = \frac{PB}{AB}$. This proportion — apparently already known to the Egyptians and the Greeks — was probably first named the "Golden Section" or "sectio aurea" by Leonardo da Vinci, who drew geometric diagrams for Fra Luca Pacioli's book, *De Divina*

*Proportione* (1509), which dealt with this topic. One of da Vinci's contributions to this book is the famous drawing of "the Vitruvian Man" as shown in Figure 3.5.

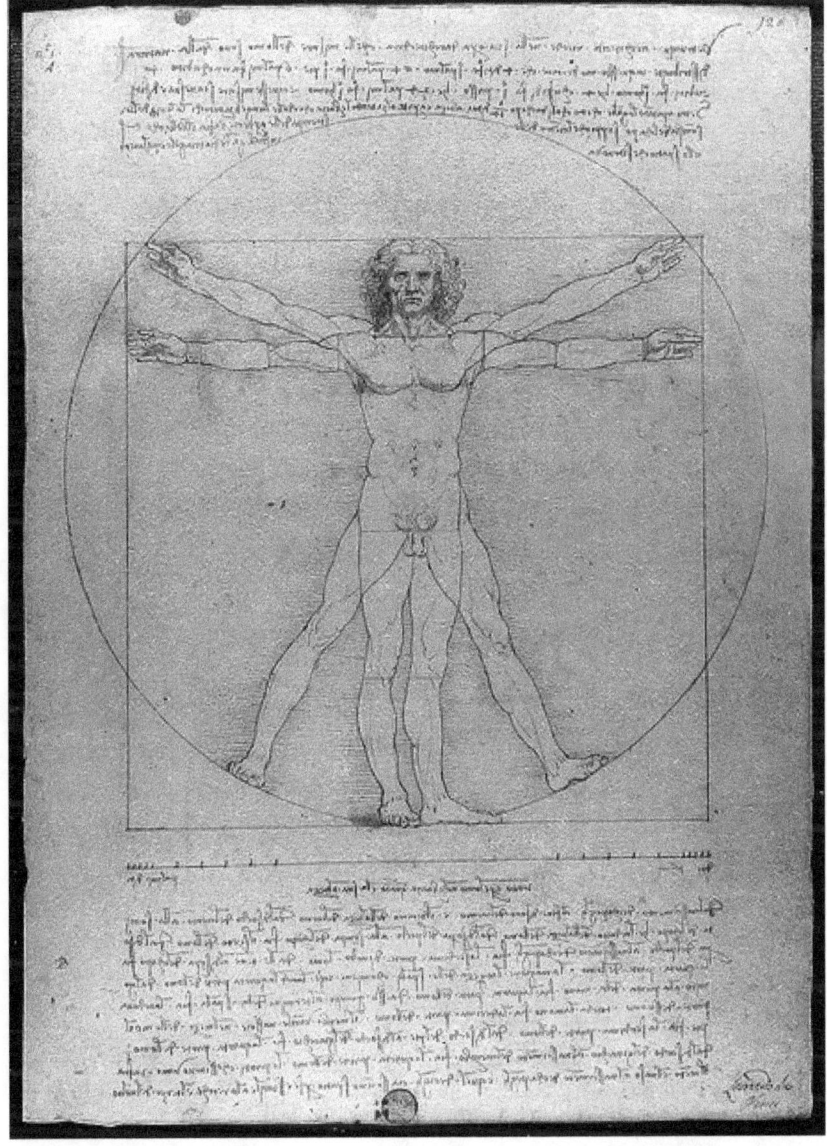

**Figure 3.5**

Da Vinci provided notes based on the work of Vitruvius (ca. 84 BCE–ca. 27).[1] The drawing, which is in the possession of the Gallerie dell'Accademia in Venice, Italy, is often considered one of the early breakthroughs of pictorially depicting a perfectly proportioned human body. Apparently, da Vinci derived these geometric proportions from Vitruvius's treatise *De Architectura*, Book III. The drawing shows a male figure in two superimposed positions with his arms and legs apart and inscribed in a circle and square, which are tangent at only one point. The Golden Ratio is exhibited in that the distance from the soles of the man's feet to his navel (which appears to be at the center of the circle, as shown in Figure 3.5) divided by the distance from the navel to the top of his head, which is about 0.656, approximates the Golden Ratio, which we know is about 0.618... .

Had the square's upper vertices been somewhat closer to the circle, then the Golden Ratio would have been attained. This can be seen in Figure 3.6, where the radius of the circle is selected to be 1, and the side of the square is 1.618, approximately equal to $\phi$, which is the symbol traditionally used to represent the Golden Ratio.

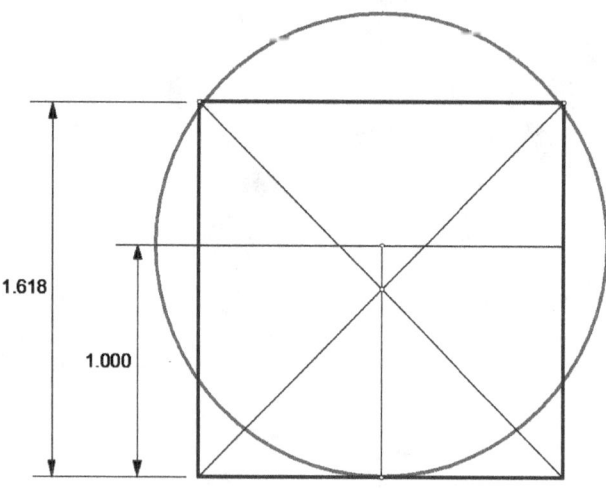

**Figure 3.6**

# The Surprising Golden Ratio

There are probably endless beauties involving this Golden Ratio. One of these is the relative ease with which one can construct the ratio by merely folding a strip of paper.

Simply take a strip of paper, say about 1″ – 2″ wide and make a knot. Then very carefully flatten the knot as shown in Figure 3.7. Notice the resulting shape appears to be a regular pentagon, that is, a pentagon with all angles congruent and all sides the same length.

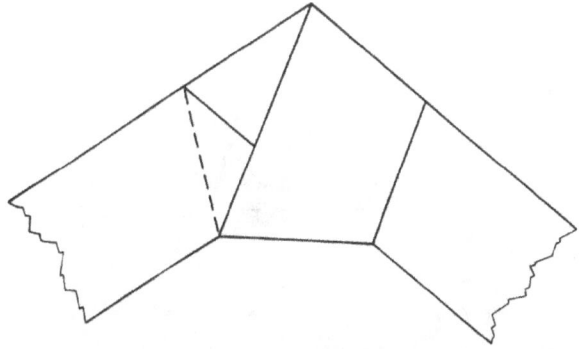

**Figure 3.7**

If you use relatively thin translucent paper and hold it up to a light, you ought to be able to see a pentagon with it diagonals. These diagonals intersect each other in the Golden Ratio as shown in Figure 3.8.

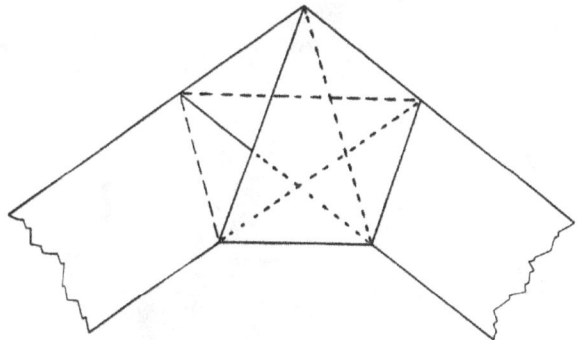

**Figure 3.8**

Let's take a closer look at this pentagon, which we show high-lighted in Figure 3.9. Point $D$ divides segment $AC$ into the Golden ratio,[2] which is $\frac{DC}{AD} = \frac{AD}{AC}$. We can also say that the segment of length $AD$ is the mean proportional between the lengths of the shorter segment ($DC$) and the entire segment ($AC$).

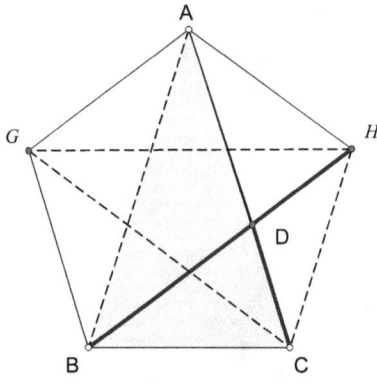

**Figure 3.9**

For some, it might be useful to show what the actual value of the Golden Ratio is. To do this, we begin with the isosceles triangle $ABC$, whose vertex angle has measure 36°. Then consider the bisector $BD$ of angle $ABC$, which is shown in Figure 3.10.

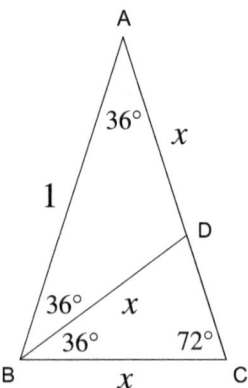

**Figure 3.10**

We find that $m\angle DBC = 36°$. Therefore $\triangle ABC \sim \triangle BCD$. Let $AD = x$ and $AB = 1$. However, since $\triangle ADB$ and $\triangle DBC$ are isosceles, $BC = BD = AD = x$. From the similarity above: $\frac{1-x}{x} = \frac{x}{1}$. This gives us $x^2 + x - 1 = 0$, and therefore, $x = \frac{1+\sqrt{5}}{2}$. (The negative root cannot be used for the length of a segment.) We, therefore, call triangle $ABC$ a *golden triangle*. The Golden ratio can be seen as the most beautiful ratio in geometry. When a rectangle's dimensions reflect that ratio, it can be seen in some of the most famous structures in the world such as the Parthenon in Athens, Greece. To further investigate this amazing ratio, see A. S. Posamentier, and I. Lehmann, *The Glorious Golden Ratio*, Amherst, New York: Prometheus Books, 2012.

# A Surprise Appearance of an Unexpected Parallelogram

There are some simple geometric occurrences that can be drawn on the paper napkin in a restaurant, when you would like to impress someone with a geometric surprise that will get them thinking and wondering why they have never seen something like that before. Begin, by having your audience draw *any* quadrilateral with no equal or parallel sides. In other words, have them draw an "ugly" quadrilateral. It is always best to have the members of the audience draw differently shaped quadrilaterals to better dramatize the ensuing result. Once each of the audience members has drawn a quadrilateral, have them locate the midpoints of each of the 4 sides of the quadrilateral and then have them connect these midpoints consecutively. Drawn properly, each person should end up with a parallelogram, as you can see with our example in Figure 3.11. They should be amazed that no matter what shape quadrilateral they began with, the end result will always be a parallelogram inside the original quadrilateral.

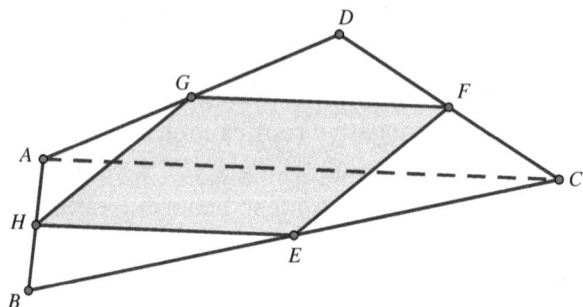

**Figure 3.11**

Justifying this result is merely the application of a theorem, which is part of the usual high school curriculum, namely, that in any triangle a line segment joining midpoints of two sides of the triangle is parallel to the third side of the triangle and one-half its length. If we draw one diagonal, say, *AC* in this original quadrilateral (Figure 3.11), we will have formed two triangles (*ADC* and *ABC*). In triangle *ABC*, we then have *HE* parallel to and one-half the length of *AC*; and in triangle *ADC* we have *GF* parallel to and one-half the length of *AC*. Therefore, we have *GF* parallel to *HD* and *GF* is equal to *HE*. Thus, we can show that we will have found that the quadrilateral formed by joining the consecutive midpoints of the quadrilateral will have two sides (*GF* and *HE*) equal in length and parallel, since they are both parallel and half the length of *AC*, thus, making the newly formed quadrilateral a parallelogram.

To further challenge your audience, you might want to ask them what characteristics must the original quadrilateral have in order for the parallelogram formed by joining the midpoints of its sides would be a rhombus, a rectangle, or a square. We show these three results in Figures 3.12–3.14.

In Figure 3.12, the diagonals of the original quadrilateral, *AC* and *DB*, are equal, and the resulting figure by joining midpoints of the quadrilateral is a rhombus.

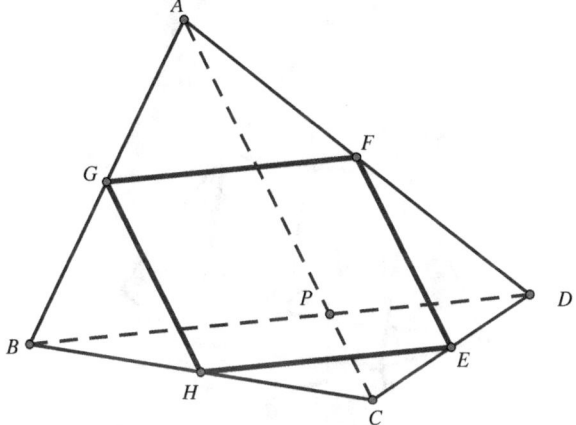

**Figure 3.12**

When the two diagonals of the original quadrilateral are perpendicular as is the case in Figure 3.13, the resulting figure *GFEH* is a rectangle.

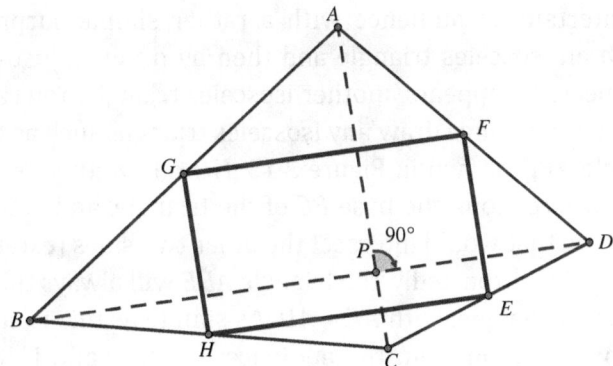

**Figure 3.13**

In Figure 3.14, when the diagonals of the original quadrilateral are equal and perpendicular, the resulting figure by joining midpoints of the original quadrilateral is a square.

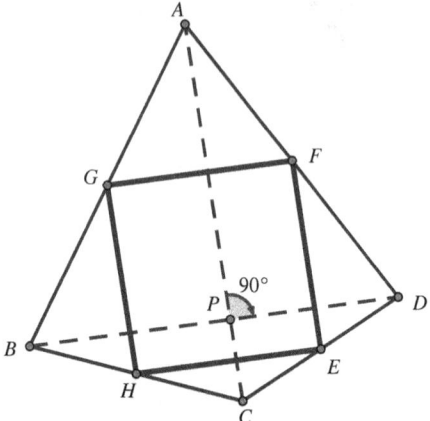

**Figure 3.14**

## The Unexpected Isosceles Triangle

You can entertain an audience with a rather simple surprise that begins with an isosceles triangle and then by drawing just one line there unexpectedly appears another isosceles triangle. You can begin by having your audience draw any isosceles triangle, such as the isosceles triangle *ABC*, shown in Figure 3.15. Next, have them select any point *P* anywhere along the base *BC* of the triangle, and erect a perpendicular line that would intersect the other two sides (extended) at points *D* and *E*. Unexpectedly, the triangle *ADE* will always turn out to be an isosceles triangle, with $AE = AD$. As simple as this is, it usually gets a "wow" reaction from the audience — especially, if it is presented in a somewhat animated or dramatic fashion.

The reason why triangle *ADE* is isosceles is because it can be easily shown that $\angle AED = \angle ADE$, since both angles are supplementary to the equal base angles of the original isosceles triangle.

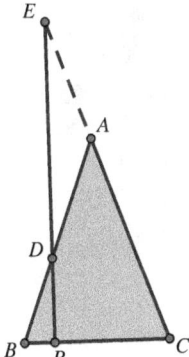

**Figure 3.15**

## Surprising Equality of Line Segments

It is well known that the angle bisectors of any triangle meet at a common point. Some might recall from their high school geometry that this point is the center of the inscribed circle of the triangle. However, this point of intersection, often called the in-center, leads us to a very surprising equality of line segments. Let's consider triangle *ABC* with its three angle bisectors meeting at point *P*, as shown in Figure 3.16. Through point *P*, the line segment *DE* is drawn, meeting sides *AB* and *AC* at points *D* and *E*, respectively. Through point *D*, the line segment *DG* is drawn parallel to *AC* meeting *BC* at point *G*. Similarly, the line segment *EF* is drawn parallel to *AB* meeting *BC* at point *F*.

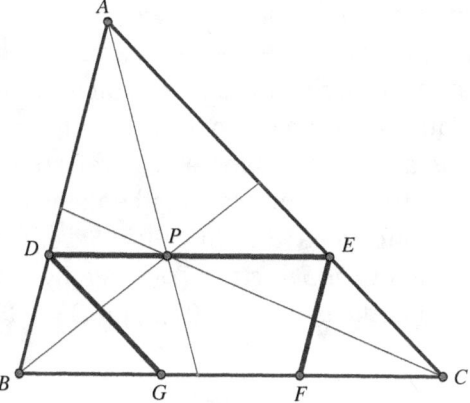

**Figure 3.16**

Completely unanticipated, we find that the sum of the lengths of $DG + EF = DE$. For the meticulous reader, this can be justified by noting that $DB = DP$, and $EP = EC$, since they are the legs of two isosceles triangles, which were so established since the base angles are easily shown to be equal. From the two parallelograms, $DECG$ and $DEFB$, we can get $DG = EC$ and $EF = DB$. Since $DP + PE = DE$, by various substitutions we can show that $DG + EF = DE$.

## The Surprise Equilateral Triangle

One of the most unusual occurrences in mathematics can also be rather entertaining. However, it requires having your audience trisect an angle. It is well known that it is not possible to trisect a general angle with the normal construction tools: an unmarked straight edge and a pair of compasses. Therefore, your audience will either have to estimate the trisection of an angle or perhaps use a protractor, or possibly have access to a dynamic drawing program on a laptop computer or tablet. In any case, you would begin by having your audience draw any scalene triangle, which would have no special properties. As you guide them through this construction, they will see that the angle trisectors of any triangle that they draw will determine an unexpected equilateral triangle. We show this phenomenon in Figure 3.17, where we have various triangles of different shapes and in each case we have drawn the trisectors of its angles. We mark the intersections of adjacent trisectors as points $D$, $E$, and $F$. In each case, the triangle formed by these three intersection points *always* determines an equilateral triangle. It is important that dramatically presenting this unexpected result is essential to properly amaze the audience. Please be aware that the justification, which can be done with nothing more than high school geometry, is a bit challenging and is referred to as Morley's theorem, attributed to the American mathematician Frank Morley (1860–1937), who published this in 1900. Several proofs of this theorem can be found in *The Secrets of Triangles*, by A. S. Posamentier and I. Lehmann (Prometheus Books, 2012), pp. 351–355.

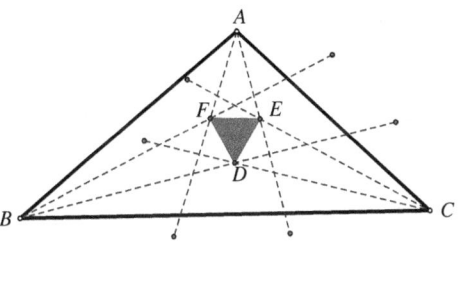

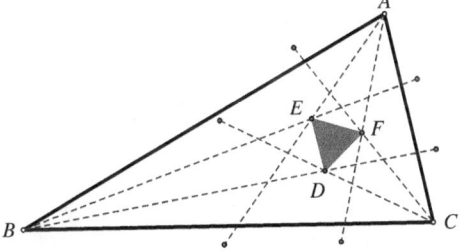

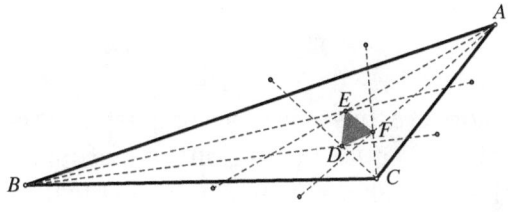

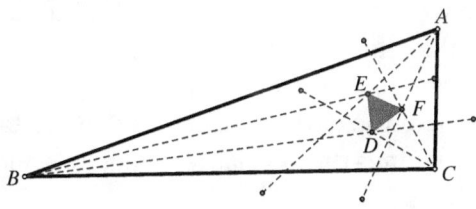

**Figure 3.17**

## Trisecting a Circle

When one is asked to divide a circle into 3 equal parts, the natural result is to draw the radii and angles 120° from each other, as we show in Figure 3.18.

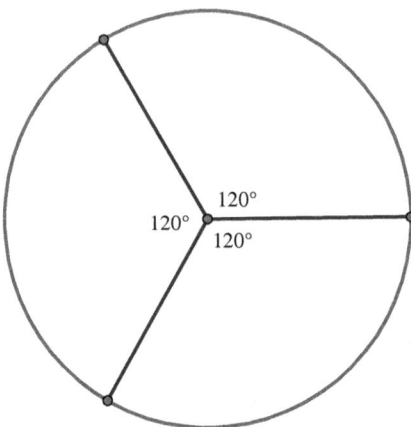

**Figure 3.18**

However, a more creative way of accomplishing this task of partitioning a circle of the 3 equal area regions can be done with the array of semicircles shown in Figure 3.19. We can justify this in the following way: We first seek the area of the darker-shaded region in Figure 3.19. Once we have that and show that it is one-third, $\frac{1}{3}$, of the area of the circle, then the similar figure at the top will also be $\frac{1}{3}$, leaving the center portion to be the remaining $\frac{1}{3}$.

To find the area of the darker-shaded region, we take the area of the large semicircle minus the next larger semicircle and after that the smallest semicircle as follows:

Area of the semicircle with radius $AB = \frac{\pi}{2}\left(\frac{r}{3}\right)^2 = \frac{\pi r^2}{18}$.

Area of the semicircle with radius $AD = \frac{\pi r^2}{2}$.

Area of the semicircle with radius $CE = \frac{\pi}{2}\left(\frac{2r}{3}\right)^2 = \frac{2\pi r^2}{9}$.

Therefore, the area of the darkly shaded region is $\frac{\pi r^2}{18} + \frac{\pi r^2}{2} - \frac{2\pi r^2}{9} = \frac{\pi r^2}{3}$, which is one-third of the area of the large circle. As we mentioned earlier, the same area can be calculated for the similar shape above. These two areas then account for two-thirds of the area of the circle leaving the center section as the remaining one-third of the circle. Thus, we have divided the circle to three equal area regions. Not only is this enlightening but it is also entertaining and demonstrating to the audience that it often pays to consider "thinking out of the box."

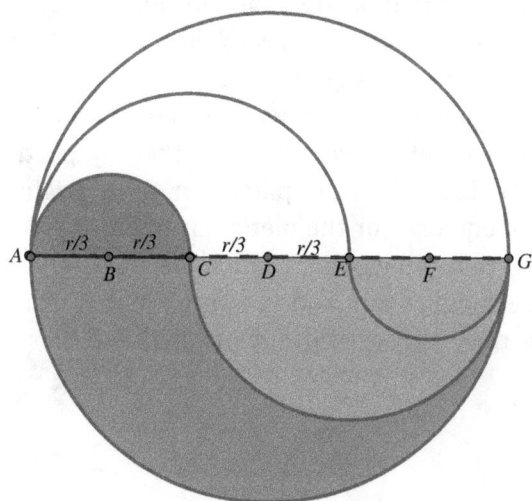

**Figure 3.19**

## Dividing Five Circles

Here we are given 5 circles tangent to each other and placed as shown in Figure 3.20. The problem here is to find a line through the center point of the leftmost circle that will equally divide the areas of these 5 circles.

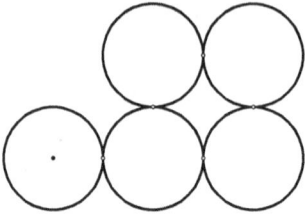

**Figure 3.20**

Without an auxiliary line, this problem would be difficult to solve. We will also add 3 additional circles to the diagram (Figure 3.21) to make our solution more easily reached. We now draw this auxiliary line through the center of the lower left circle and containing the center of the upper right circle. We have, thereby, partitioned these circles into 2 equal area parts. In particular, the 5 original circles are now partitioned equally. For the meticulous reader, we offer the fact that the angle $\alpha$ that the auxiliary line makes with the horizontal is an angle whose tangent is $\frac{1}{3}$ and is approximately equal to 18.43°. Thus, we have then partitioned our original 5 circles into 2 equal areas.

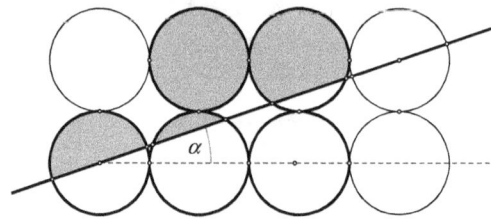

**Figure 3.21**

## Pick's Theorem

Here we will take the audience out of the traditional realm of the common geometry. The areas of rectangles and triangles are familiar, but what is the area of something more exotic-looking like the shaded polygon shown in Figure 3.22?

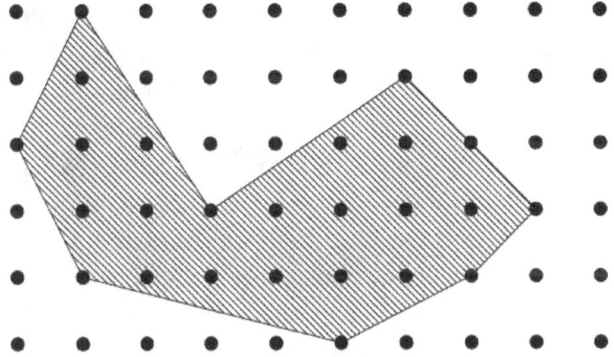

**Figure 3.22**

When studying geometry and area, it is customary to first understand the areas of basic shapes such as triangles, rectangles, and circles. For the areas of more complicated shapes like the one shown in
Figure 3.22, a standard practice is to cut up the figure into more manageable smaller basic pieces, then add the areas of the smaller pieces
together to obtain the area of the whole.

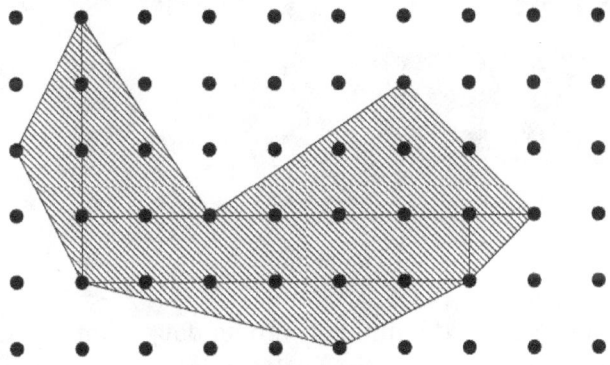

**Figure 3.23**

Figure 3.23 illustrates this simplifying procedure to reduce the
problem into one of counting the areas of triangles and rectangles. It
is an exercise to calculate the areas of the 5 triangles and 1 rectangle
shown, then add it all up to get a total area of 19.5 square units. This
process will certainly work in this case to give us the shaded area, but

there is another much simpler method available called Pick's theorem, named after Austrian mathematician Georg A. Pick (1859–1942), who discovered it in 1899. With this you will truly entertain your audience.

The dots in the diagrams above are *lattice points*, that is, points in the plane whose *x*- and *y*-coordinates are both integers. The *x*- and *y*-axes are not important here, so they are omitted in the diagram. A *lattice polygon* is a polygon whose vertices are lattice points. The shaded polygon in Figure 3.22 is an example of a lattice polygon.

Pick's Theorem gives a simple formula to compute the area of a lattice polygon by counting points in the polygon. The *boundary points*, as the name suggests, are the lattice points on the boundary of the lattice polygon. The boundary points are circled in Figure 3.24. How many boundary points do you see?

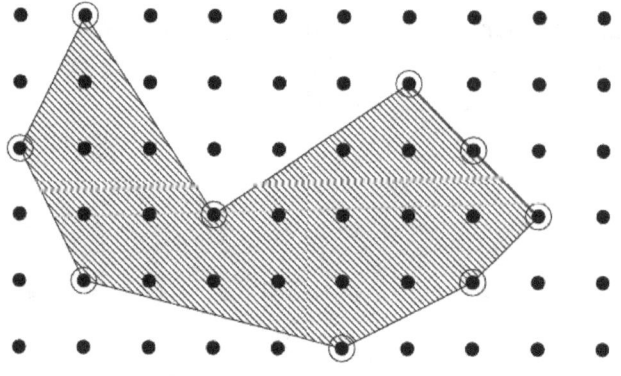

**Figure 3.24**

Let's define *B* to be the number of boundary points. In this case, *B* = 9. The *interior points* are the lattice points contained inside the lattice polygon, but not on the boundary itself. The interior points are circled in Figure 3.25. How many interior points are there?

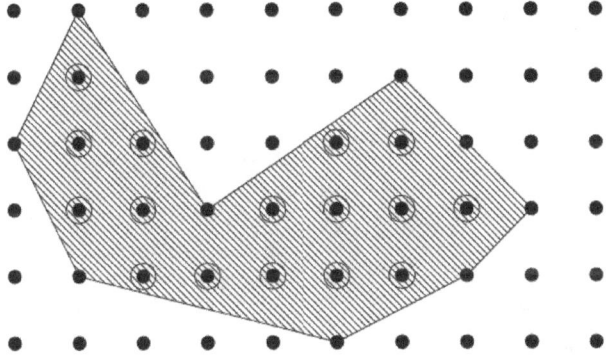

**Figure 3.25**

Let's define $I$ to be the number of interior points. In this case, $I = 16$. Pick's theorem states that the area $A$ of a lattice polygon can be computed as follows: $A = \frac{B}{2} + I - 1$. In our example above, the area is $A = \frac{9}{2} + 16 - 1 = 19.5$.

When faced with a region that's more complicated than the basic shapes we are accustomed to, it is good practice to simplify the situation by decomposing the region into basic shapes. This strategy of reducing the complex to the more familiar and simple permeates much of mathematics. Pick's theorem carries this philosophy even further than one might suspect possible, reducing the problem of calculating the areas of lattice polygons into one of merely counting dots. To drive this method home, it would be wise to allow your audience to try Pick's theorem with other strange looking polygons on a lattice plane.

## Cutting Up Fiver Squares to Make One Square

There are times when it can be entertaining to challenge your audience geometrically. Here we are asking them to make 2 cuts of the figures as shown in Figure 3.26 and place them in such a way that one square results.

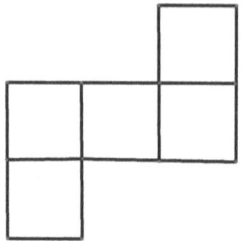

**Figure 3.26**

After you've had a chance to allow your audience to ponder this challenge question you might show them to cuts, which we show in Figure 3.27.

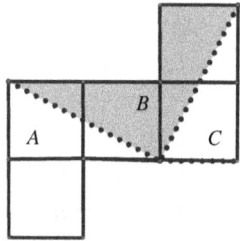

**Figure 3.27**

The trick for them is then to see how these two cuts can be rearranged to form a square. We show this in Figure 3.28. You will notice that the side of the new square has length $\sqrt{5}$, since if we consider triangle C and apply the Pythagorean theorem, we get: $1^2 + 2^2 = S^2$; therefore, $S = \sqrt{5}$.

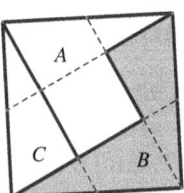

**Figure 3.28**

# A Slice in Half

A challenge that is not so obvious can be entertaining by dint of the fact that the solution will be surprising and yet not difficult to understand. Such is the case with the conundrum that we propose in Figure 3.29, where all the 13 squares shown are equal area and the challenge is to make one cut through point *X* so that the total area of the two parts will be equal.

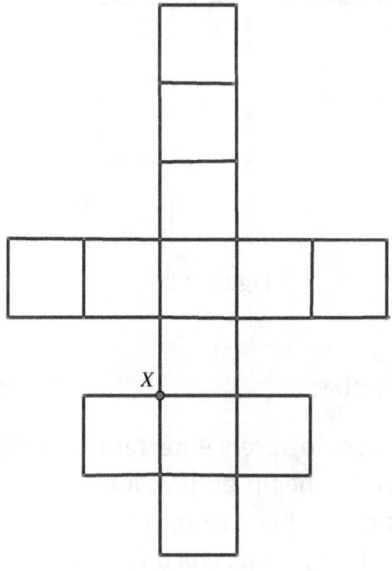

**Figure 3.29**

We see that the area of the entire figure consists of 13 squares. Therefore, we need to indicate a cut will divide this entire figure into $6\frac{1}{2}$ or 6.5 squares. We begin by selecting a point *Y* at the midpoint square as shown in Figure 3.30. This allows us to create the shaded rectangle *XZYW*, whose area is 1.5 × 2 = 3. The diagonal *XY* splits this rectangle in half so consequently the triangle *XWY* has an area of 1.5, which when added to the 5 squares above and to the left of line *XY* result in an area of 6.5, which is half the total area of the original 13 square figure.

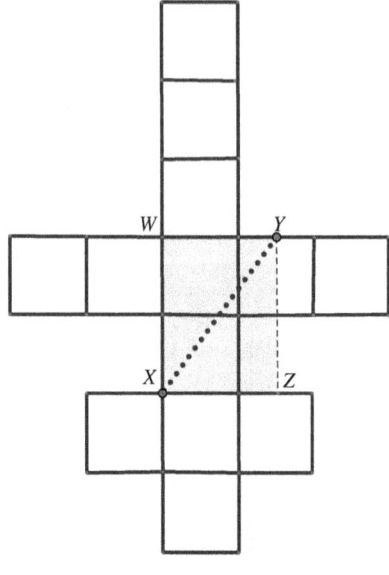

**Figure 3.30**

## Cutting Up a Circle

With this activity you could surely entertain an audience to search for a desired conclusion. We begin by providing a circle and asking the audience to use 6 straight lines to partition the circle into as many sections as possible. Clearly, some will do what is shown in Figure 3.31, where the 6 lines partition the circle into 16 sections. This is clearly not the maximum number of sections that can be partitioned by 6 straight lines.

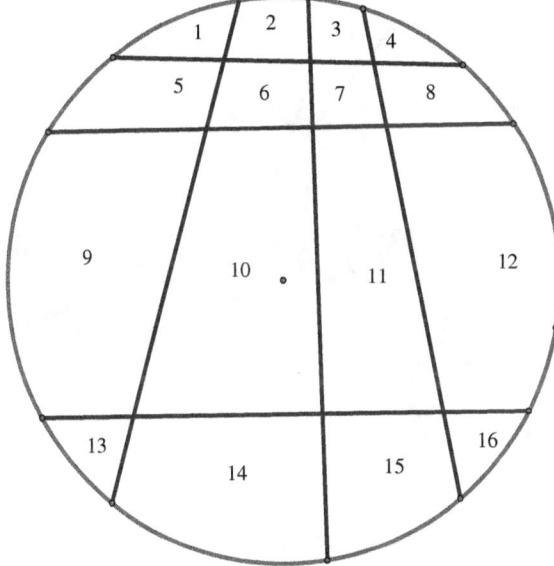

**Figure 3.31**

When you inform your audience that this is not by far the most number of sections that the 6 straight lines could produce, they will begin to investigate further. It may take some time till they realize that each line they draw must intersect every one of the other 5 lines and no 3 lines should ever be concurrent, that is, there should be no intersection point containing more than 2 lines. Eventually, they should be able to produce 22 partitions of the circle similar to that shown in Figure 3.32.

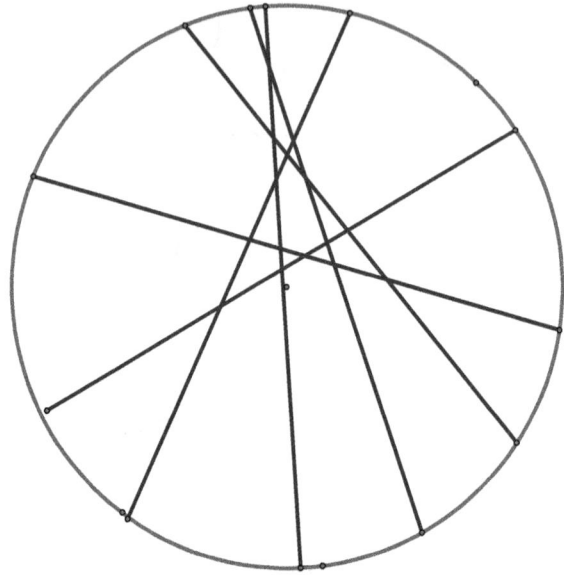

**Figure 3.32**

# A Conundrum of Probability in Geometry

With this seemingly simple situation, you may leave your audience in a rather upsetting tone because of its counterintuitive nature. There are times when seemingly simple issues can be stretched to producing a conundrum that is not easily rectified. One such example is with two concentric circles, where the radius of the smaller circle is one-half the radius of the larger circle as shown in Figure 3.33. The question is: What is the probability that a point selected in the larger circle is also located in the smaller one? The typical (and correct) answer is $\frac{1}{4}$. This can easily be shown by letting a smaller circle's radius, $OA$, be represented by $r$, and the large circle's radius be represented by $R$, where $r = \frac{1}{2}R$. Then the area small circle is $\pi \left(\frac{1}{2}R\right)^2 = \frac{1}{4}\pi R^2$, that is, one-fourth of the area of the larger circle, which is $\pi R^2$.

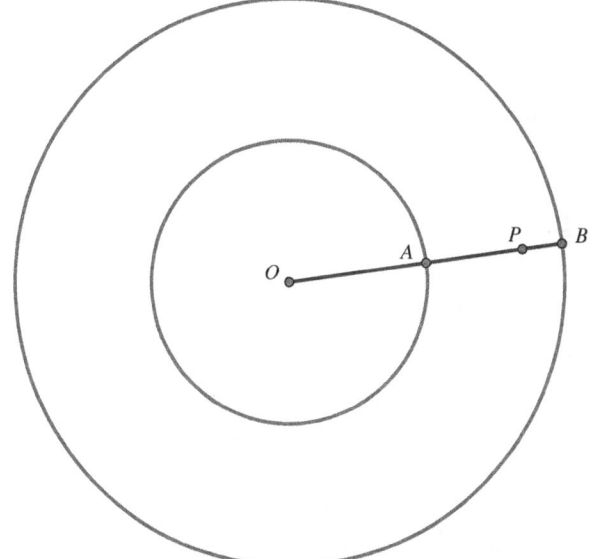

**Figure 3.33**

Therefore, if a point is selected at random in the larger circle, the probability that it would be in the smaller circle as well is $\frac{1}{4}$.

Now here is where the conundrum comes in, when we look at this question differently. The randomly selected point $P$ must lie on some radius line of the larger circle. Let's say we will use radius $OAB$, where $A$ is its midpoint. The probability that a point $P$ on $OAB$ would be on $AB$ is $\frac{1}{2}$, since $AB = \frac{1}{2}OB$. Now, if we were to do this for any other point in the larger circle, we would find the probability of the point being on a radius line and in the smaller circle is $\frac{1}{2}$. This, of course, is *not* correct, although it seems perfectly logical. Where is the error? Here is where you expose the background, which is actually an attempt to explain a conundrum. The "error" lies in the initial definition of each of two different sample spaces, that is, the set of possible outcomes of an experiment. In the first case, the sample space is the entire area of the larger circle, while in the second case, the sample space is the set of points on a radius such as $OAB$. Clearly, when a point is selected on $OAB$, the probability that the point will be on $AB$ is $\frac{1}{2}$. These are two entirely different problems even though (to dramatize the issue) they appear to be the same. Conditional probability is an important concept to stress, and what better way to instill this idea than through a demonstration that shows obvious absurdities. Perhaps this conundrum will motivate the audience to investigate this further.

## More Geometric Conundrums

Suppose we draw a random chord in a circle circumscribed around an equilateral triangle. The question with which we are faced is: What is the probability that the chord drawn will be longer than a side of the triangle. We see in Figure 3.34, where the required chord could be drawn through point $A$ of triangle $ABC$. In that case, the chord could be either $AP$, $AR$, or $AQ$. The point $Q$ could be anywhere on arc $BC$. In that case, the chord would be longer than the length of the side of the triangle, whereas if it landed anywhere on arc $AB$ or arc $AC$, it would be shorter than the side of the triangle. Since arc $BC$ is $\frac{1}{3}$ of the circle, the probability that the chord is longer than the length of the side is $\frac{1}{3}$.

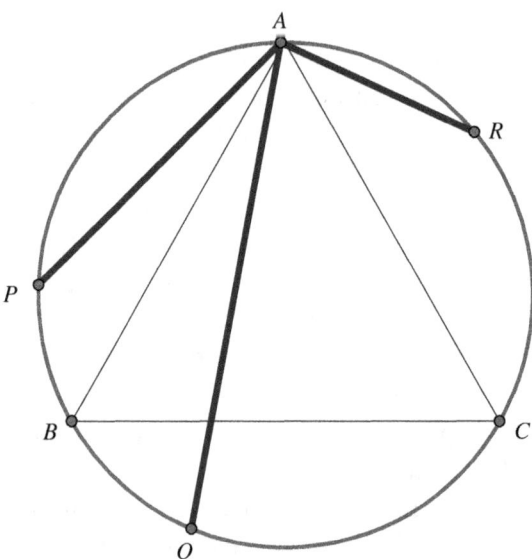

**Figure 3.34**

Now let's consider that the chord is located on arc *AB*, but not sharing the vertex point *A* as we had in Figure 3.34. We show this new position in Figure 3.35. Here there are two possibilities for drawing the chord. Chord *PS* could be drawn so that it does not intersect the equilateral triangle, or the chord could be drawn as we show with *RQ*, which does intersect the triangle.

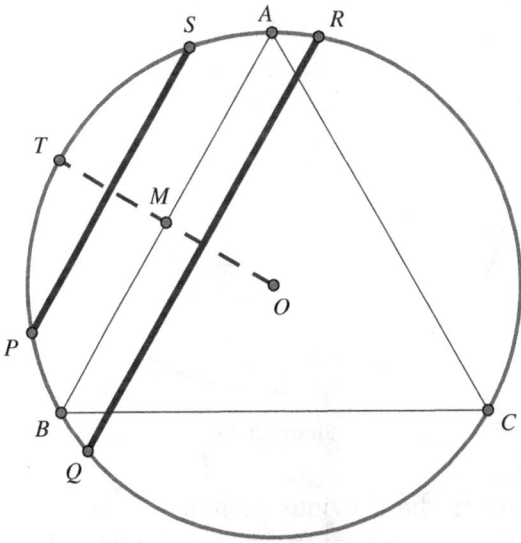

**Figure 3.35**

Before we move any further, we need to digress a bit to show that the intersection of a radius perpendicular to the side of a triangle, when the chord is parallel to that side, is the midpoint of the radius. In Figure 3.36, we had drawn some auxiliary lines, where we can show that the quadrilateral *AOBT* is a parallelogram[3] where the diagonals bisect each other. This establishes that *M* is the midpoint of *OT*. The chord drawn could either intersect the perpendicular radius between points *T* and *M*, or between points *M* and *O*. The chord will be longer than the side of the triangle if it intersects the perpendicular radius between points *M* and *O*. The chances for that to happen is equal to the chance that it will intersect between points *T* and *M*.

Therefore, the probability that the line will be longer than the side of the equilateral triangle is $\frac{1}{2}$.

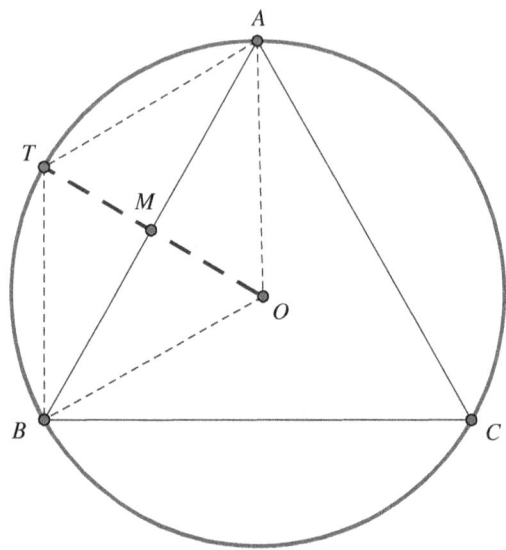

**Figure 3.36**

This contradicts the previous response. However, if you revert back to the earlier example of the radius shown in Figure 3.33, you will have some explanation for this awkward result.

There is still another variation to this situation. Suppose we consider a circle inscribed in the equilateral triangle as shown in Figure 3.37. We know that the area of the circle inscribed in triangle *ABC* is $\frac{1}{4}$ of the area of the circumscribed circle, since the radius of the inscribed circle is $\frac{1}{2}$ the length of the circumscribed circle. If the chord in question intersects the inscribed circle, then its length is longer than the side of the triangle *ABC*. Another way of saying this is that the midpoint *M* of the chord *QR* would have to be within the inscribed circle. The probability that midpoint *M* would lie in the inscribed circle is $\frac{1}{4}$ and the probability here that the length of the chord is longer than the side of the triangle is also $\frac{1}{4}$.

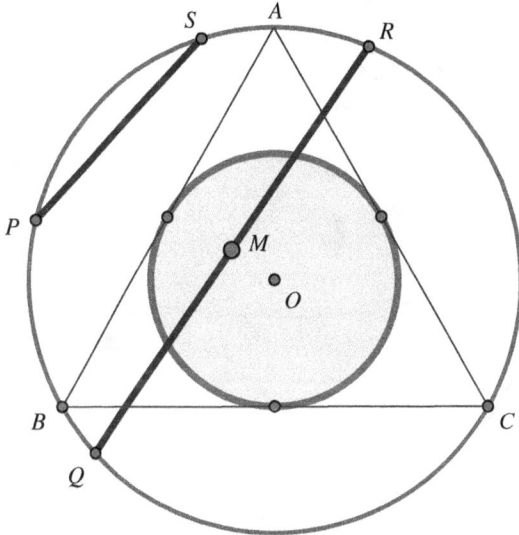

**Figure 3.37**

The curiosity here is that we have come up with three different likelihoods or probabilities for drawing a randomly selected chord that is longer than a side of the equilateral triangle. This entertainment would certainly merit further discussion, which it is intended to do.

## Where in The World Are You?

There are entertainments in mathematics that stretch (gently, of course) the mind in a very pleasant and satisfying way. This unit presents just such a situation. We will begin with a popular puzzle question that has some very interesting extensions, which are seldom considered (but we will consider them later on). It requires some "out of the box" thinking that leave you with some favorable lasting effects. Let's consider the question:

*Where on earth can you be so that you can walk: <u>one mile south</u>, then <u>one mile east</u>, and then <u>one mile north</u> and end up at the starting point?* (see Figure 3.38 not drawn to scale, obviously!).

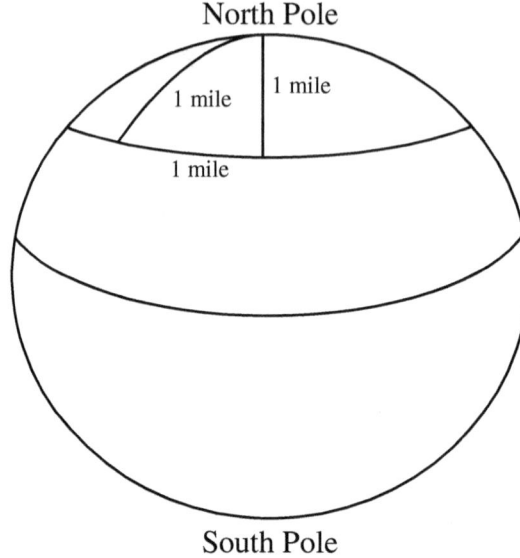

North Pole

South Pole

**Figure 3.38**

Mostly through trial and error, a clever audience will stumble on the right answer: The North Pole. To test this answer, try starting from the North Pole and traveling south one mile and then east one mile takes you along a latitudinal line which remains equidistant from the North Pole, one mile from it. Then traveling one mile north gets you back to where you began, the North Pole.

Most people familiar with this problem feel a sense of completion. Yet we can ask: Are there other such starting points, where we can take the same three "walks" and end up at the starting point? The answer, surprising enough is *yes*.

One set of starting points is found by locating the latitudinal circle, which has a circumference of one mile, and is nearest the South Pole. From this circle walk one mile north (along a great circle, naturally), and for another latitudinal circle. Any point along this second latitudinal circle will qualify. Let's try it. (see Figure 3.39 not drawn to scale, obviously!).

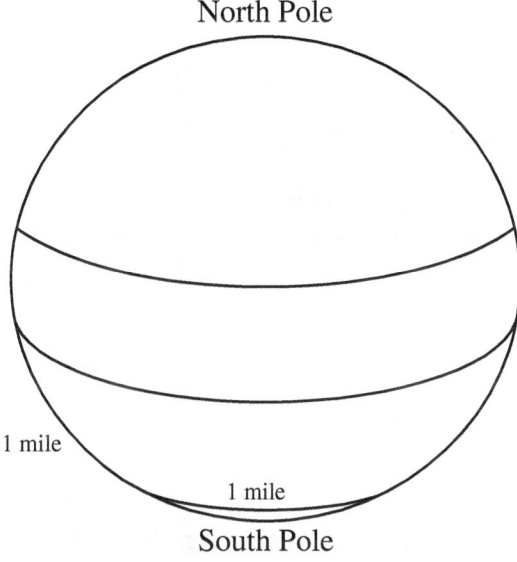

North Pole

1 mile

1 mile

South Pole

**Figure 3.39**

Begin on this second latitudinal circle (the one further north). Walk one mile south (takes you to the first latitudinal circle), then one mile east (takes you exactly once around the circle), and then one mile north (takes you back to the starting point).

Suppose the second latitudinal circle, the one we would walk along, would have a circumference of $\frac{1}{2}$ mile. We could still satisfy the given instructions, yet this time walking around the circle *twice*, and get back to our original starting point. If the second latitudinal circle had a circumference of $\frac{1}{4}$ mile, then we would merely have to walk around this circle *four* times to get back to the starting point of this circle and then go north one mile to the original starting point.

At this point, we can take a giant leap to a generalization that will lead us to many more points that satisfy the original stipulations. Actually an infinite number of points! This set of points can be located by beginning with the latitudinal circle, located nearest the south pole, and has a $\frac{1}{n}$th-mile circumference, so that an $n$-mile walk east will take you back to the point on the circle at which you began your walk on this latitudinal circle. The rest is the same as before, that is, walking one mile south and then later one mile north. Is this possible with latitude circle routes near the North Pole? Yes, of course!

Analogous to this issue is the well-known question about where a person could build a regular four-sided house, where each side had a southern view. Of course, the answer is the house would be built on the North Pole, where each side's exposure is in a southerly direction.

## Geometric Limits with Understanding

The concept of a limit is not to be taken lightly. It is a very sophisticated concept that can be easily misinterpreted. Sometimes the issues surrounding the concept are quite subtle. Misunderstanding of these can lead to some curious (or humorous, depending on your viewpoint) situations. This can be nicely exhibited with the following two illustrations. Assure your audience in advance not to be upset with the end results as they could be somewhat upsetting. Remember this is for entertainment. Consider the following two illustrations separately and then they should be able to notice their connection.

In Figure 3.40, it is easy to see that the sum of the lengths of the bold segments (the "stairs") is equal to $a + b$.

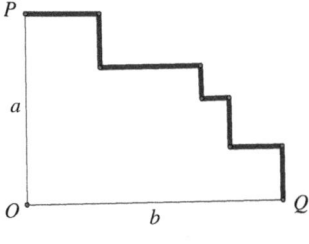

**Figure 3.40**

The sum of the bold segments ("stairs"), found by summing all the horizontal and all the vertical segments, is $a+b$. If the number of stairs increases, the sum is still $a+b$. The dilemma arises when we increase the stairs to a "limit" so that the set of stairs appears to be straight line, which in this case would be the hypotenuse of $\triangle POQ$. It would then appear that $PQ$ has length $a + b$. Yet we know from the Pythagorean theorem that $PQ = \sqrt{a^2 + b^2}$ and *not* $a + b$. So, what's wrong?

Nothing is wrong! While the set consisting of the increasing number of stairs does indeed approach closer and closer to the straight line segment $PQ$, it does *not*, therefore, follow that the *sum* of the bold (horizontal and vertical) lengths approaches the length of $PQ$, contrary to one's intuition. There is no contradiction here, only a failure on the part of our intuition.

Another way to "explain" this dilemma is to argue the following. As the "stairs" get smaller, they increase in number. In an extreme situation, we have 0-length dimensions (for the stairs) used an infinite number of times, which then leads to considering $0 \cdot \infty$, which is meaningless!

A similar situation arises with the following example.

In Figure 3.41, the smaller semicircles extend from one end of the large semicircle's diameter to the other.

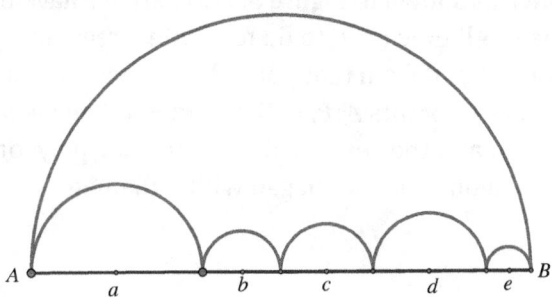

**Figure 3.41**

It is easy to show that the sum of the semicircular arc lengths of the smaller semicircles is equal to the arc length of the larger semicircle. That is, the sum of the smaller semicircles is equal to $\frac{\pi a}{2} + \frac{\pi b}{2} + \frac{\pi c}{2} + \frac{\pi d}{2} + \frac{\pi e}{2} = \frac{\pi}{2}(a+b+c+d+e) = \frac{\pi}{2}(AB)$, which is the arc length of the larger semicircle. This may not "appear" to be true, but it is! As a matter of fact, as we increase the number of smaller semicircles (where, of course, they get smaller), the sum "appears" to be approaching the length of the segment *AB*, but, in fact, does not!

Again, the set consisting of the semicircles does indeed approach the length of the straight-line segment *AB*. It does *not* follow, however, that the *sum* of the semicircles approaches the *length* of the limit, in this case *AB*.

This "apparent limit sum" is absurd, since the shortest distance between points *A* and *B* is the length of segment *AB*, not the semicircle arc *AB* (which equals the sum of the smaller semicircles). This is an important concept of limits and may be best explained with the help of these motivating illustrations, so that future misinterpretations can be avoided.

## The Surprising Appearance of a Regular Polygon

Your audience can have some fun in drawing a regular polygon by merely constructing perpendiculars in a setup circle with diameters. Merely present them with a circle with any number of equally spaced diameters drawn as shown in Figure 3.42. There we have 6 diameters drawn in circle *O*. All they need to do to create a regular polygon is to choose any point *P*, and from that point draw a perpendicular to each of the 6 diameters — points *A, B, C, D, E,* and *F*. When we join the feet of the perpendiculars, the result will be a regular polygon, which in this case is a hexagon, since we began with 6 diameters.

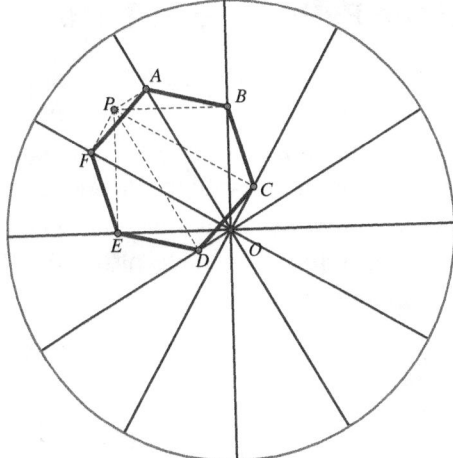

**Figure 3.42**

There is another curiosity that evolves here, namely that the center of the regular polygon constructed is located on line *PO*, which we can see by drawing a few diagonals in the regular polygon, namely, *AD* and *BE*, which intersect at point *R*, and which is also the center of the polygon and lines on *PO*, thereby, making lines *AD*, *BE*, and *PO* concurrent (see Figure 3.43).

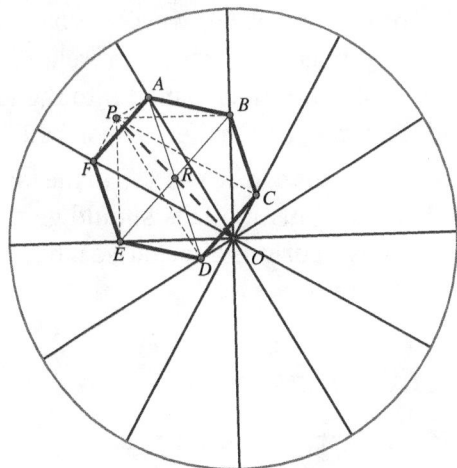

**Figure 3.43**

# A Variation of the Pythagorean Theorem

Probably, the most recalled relationships from high school geometry is the Pythagorean theorem, where $a^2 + b^2 = c^2$. We can use this relationship to create a very unusual "alternative." You can entertain your audience by using very elementary algebra and geometry to show them how to obtain the following alternative relationship: $a^{-2} + b^{-2} = c^{-2}$. We refer to Figure 3.44, where we have altitude *CD* drawn to the hypotenuse *AB* of triangle *ABC*.

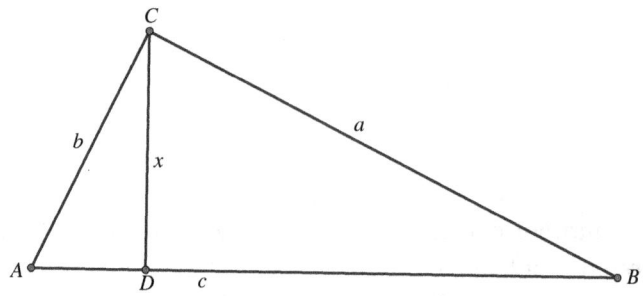

**Figure 3.44**

We can express the area of triangle *ABC* in two ways as follows: Area of $\triangle ABC = \frac{1}{2}ab = \frac{1}{2}xc$. This could also be written as $ab = xc$. By squaring both sides of this equation, we get $a^2 b^2 = x^2 c^2$. When we replace $c^2$ with the Pythagorean theorem relationship, we get $a^2 b^2 = x^2(a^2 + b^2)$, which can be transformed into the following equation: $\frac{a^2 + b^2}{a^2 b^2} = \frac{1}{x^2}$, and that can be further transformed using simple algebra to read: $\frac{1}{a^2} + \frac{1}{b^2} = \frac{1}{x^2}$. This can be written in the form $a^{-2} + b^{-2} = c^{-2}$, which is what we wanted to justify. This should give your audience something to talk about as a curious alternative when referring to the famous Pythagorean theorem.

# Equal-Area Right Triangles

It is very easy to get to right triangles that have equal areas. All you need to do is to divide a square in half by drawing a diagonal, and you will have two equal-area right triangles. However, it is not so easy to find two equal-area right triangles who sides have integer lengths.

This may be an interesting activity to present to an audience. We offer here three right triangles that have the same size area: (24, 70, 74); (40, 42, 48); and (15, 112, 133). Let's see what your audience can come up with.

## Arriving at the Unexpected

There are times in geometry when a simple situation leads to a rather unexpected and surprising result. Let's consider the quadrilateral *ABCD* inscribed in circle *O*, which has perpendicular diagonals *AC* and *BD*. We show this in Figure 3.45. Now here comes the surprising feature: when we draw a perpendicular from the point of intersection of the two perpendicular diagonals to one of the sides of the quadrilateral, that perpendicular line bisects the opposite side of the quadrilateral. In Figure 3.45, if the perpendicular from point *P* to side *AD* of the quadrilateral is extended through point *P*, it bisects the opposite side *BC* of the quadrilateral at point *M*. Of course, this can be done with the other pair of opposite sides as well. It would be helpful here when presenting this to an audience, for the presenter to show proper enthusiasm and wonder about this apparent "coincidence."

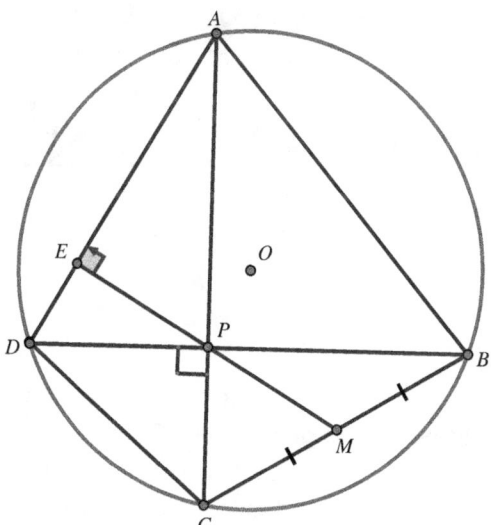

**Figure 3.45**

## Another Unexpected Result

This time we will consider a randomly drawn triangle *ABC* with two of its angle bisectors drawn: *AE* and *CD*, as shown in Figure 3.46. We then select any point *P* on the line *DE* and from that point draw perpendiculars to each of the three sides of the triangle. Now here is the amazing thing: No matter where you place point *P* on line *DE*, the sum of the distances *PN* + *PK* = *PL*. If you have available a dynamic geometry software program, then you can slide the point *P* along line *DE* and notice that the sum of the lengths of *PN* and *PK* will always be the same as the length of *PL*. Properly presented, this should surely amaze your audience.

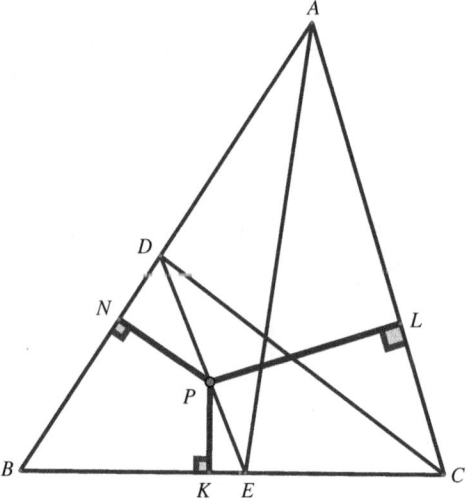

**Figure 3.46**

## The Arbelos

We credit the famous Greek mathematician Archimedes (287–212 BCE) with having discovered properties of a rather well-known geometric figure, often referred to as the *Arbelos* or *shoemakers knife*. This is shown in Figure 3.47, and is formed by three semicircles with the sum of the diameters of the two smaller ones equal to that of the larger one.

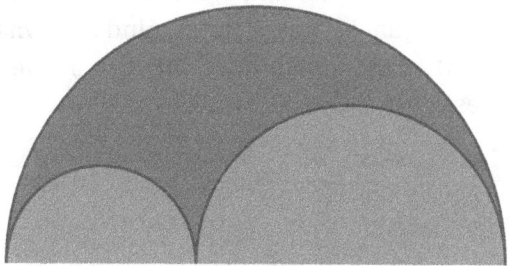

**Figure 3.47**

One of the first things to notice about this configuration is that the sum of the two smaller semicircular arcs is equal in length to the larger semicircular arc. Using the labeling shown in Figure 3.48 (the radii of the three semicircles are $AD = r_1$, $BE = r_2$, and $AO = R$), we note that the sum of the two smaller semicircular arc lengths is $\pi r_1 + \pi r_2 = \pi(r_1 + r_2) = \pi R$, which is the arc length of the larger semicircle.

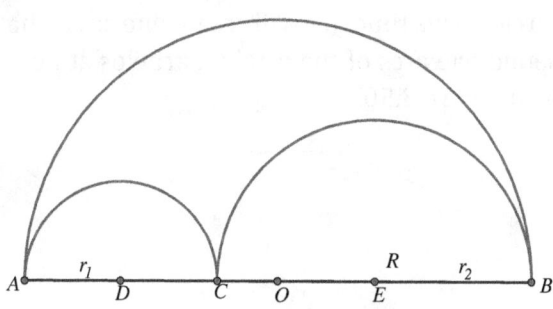

**Figure 3.48**

The arbelos presents us with a wide variety of unusual relationships, which in and of themselves can be rather entertaining even if we don't go through the trouble of justifying or proving them. For example, in Figure 3.49, we have inserted two small circles, $X$ and $Y$, which are tangent to each of the semicircular arcs and also tangent to the common tangent segment of the original two smaller semicircles. It turns out that regardless of the size of the two original smaller semicircles, centered at points *D* and *E*, the two circles, with centers *X* and *Y*, will always be equal in area. Bear in mind that what makes this so unusual is that it is independent of the size of the two original smaller semicircles.

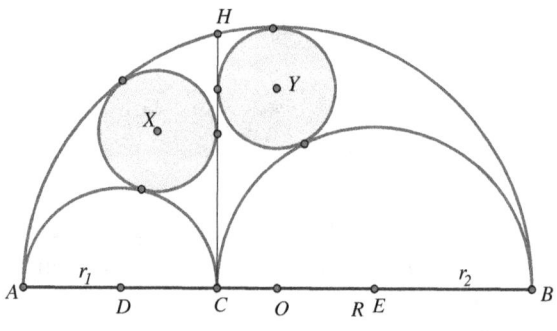

**Figure 3.49**

Now keep these two equal circles *X* and *Y* in mind as we create another circle in a most unusual fashion whose area will be the same as these two circles. This time, we will create one circle that is tangent to the three semicircle arcs of the original arbelos at points *K*, *L*, and *N*, as we show in Figure 3.50.

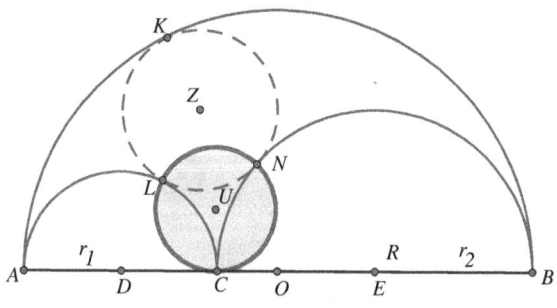

**Figure 3.50**

When we draw the circle containing the points *L*, *N*, and *C*, we find that the circle has the same area as the two circles *X* and *Y*, as shown

in Figure 3.49. Your audience should see this as a truly amazing relationships that these three circles that seem to be rather unrelated
other than being inscribed in a specific way in the same arbelos have
equal areas. Now to top that, we will come up with a fourth circle that
also shares the same area as the earlier three circles that we have so
far highlighted.

This time, we will be adding two new circular arcs. The first arc
will be drawn with center *A* and radius *AC*, and in the second arc will
be drawn with center *B* and radius *BC*. We show this with the dashed
lines in Figure 3.51. When we construct a circle, which is tangent to
each of these two newly drawn arcs and also tangent to the largest
semicircle of the arbelos, we once again have a circle equal in area to
the three previously equal circles.

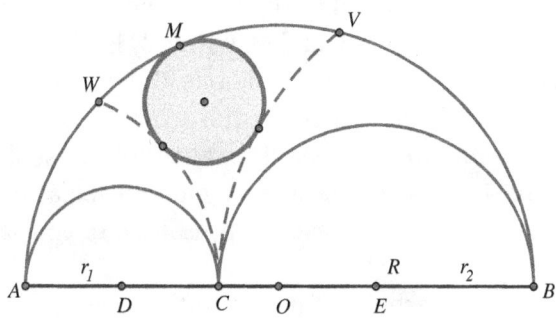

**Figure 3.51**

Although the constructions of all of these three equal-area circles
could be a bit more complicated, we offer a source that determines
the circles and the justification for their equal areas.[4]

While we have these two new circular arcs centered at *A* and *B*
and intersecting the large semicircle at points *W* and *V*, we can once
again create another pair of equal area circles, as shown in Figure 3.52.

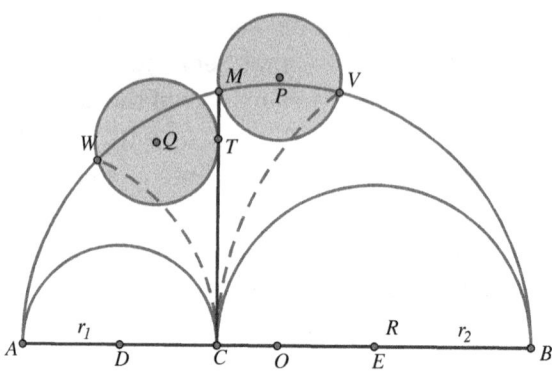

**Figure 3.52**

Circle $Q$ is the smallest circle that contains point $W$ and is tangent to $MC$, which is the perpendicular to line $AB$, and circle $P$ is the smallest circle that contains points $V$ and $M$, where $M$ is the intersection of the perpendicular $MC$ in the large semicircle. Once again, we have two equal circles essentially generated via the arbelos. As you can see so far, the arbelos seems to be an endless array of unusual geometric relationships. But it doesn't end here as you will see as you read on. Remember, the way in which these are presented to your audience is important.

Let's explore the arbelos further with some auxiliary lines. Although we will be getting a little bit more technical here than most other parts of the book, an interested audience may want to see how we draw conclusions regarding the arbelos. As shown in Figure 3.53, we will now construct a perpendicular to the line segment $AB$ through point $C$ to intersect the larger semicircle at point $H$. We then draw a common tangent to the two smaller semicircles, with tangent points $F$ and $G$, respectively, and intersecting segment $HC$ at point $S$. Finally, we draw a perpendicular from the point $D$ to radius $GE$ intersecting at point $J$.

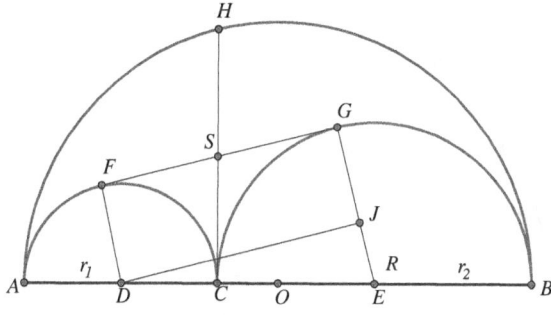

**Figure 3.53**

If we were to consider triangle *AHB*, we would notice that it is a right triangle and that the altitude to the hypotenuse, *HC*, is a mean proportional between the two segments along hypotenuse, namely, *AC* and *BC*. From that relationship, we get $HC^2 = 2r_1 \cdot 2r_2 = 4r_1r_2$. We can easily show that quadrilateral *DFGJ* is a rectangle, and, therefore, *FG* = *JD*. We also have $JE = r_2 - r_1$, and $DE = r_2 + r_1$. If we apply the Pythagorean theorem to triangle *DJE*, we get $JD^2 = (r_2 + r_1)^2 - (r_2 - r_1)^2 = 4r_1r_2$, which then has $FG^2 = 4r_1r_2$. Therefore, *HC* = *FG*, since both are equal to $4r_1r_2$. Taking this a step further, we notice that *SC* is the common internal tangent of the two smaller semicircles, so we know that *SF* = *SC* = *SG*. We can then conclude that the two segments *FG* and *HC* bisect each other and are of the same length. Therefore, a circle with center at point *S* will contain the points *F, C, G,* and *H,* which in and of itself is quite spectacular because when more than three points lie on a circle it is clearly noteworthy and to be appreciated. This must be stressed appropriately to the audience. We show this in Figure 3.54.

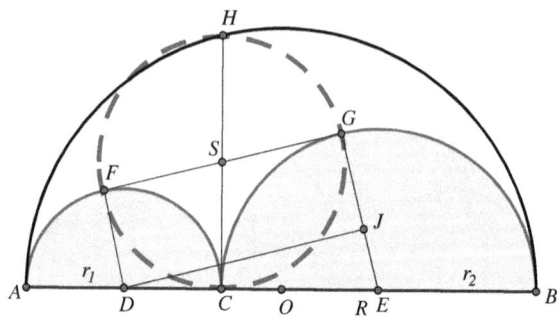

**Figure 3.54**

Now having established this unexpected circle, we can show that it has a very specific relationship to the arbelos. Amazingly, the area of the circle with center at point $S$ is equal to the area of the arbelos — which is the unshaded area between the three semicircular arcs. This can be easily justified for the more ambitious reader in the following way:

The area of the arbelos can be found by taking the area of the largest semicircle and subtracting the areas of the two smaller semicircles as follows: $\frac{\pi R^2}{2} - \left(\frac{\pi r_1^2}{2} + \frac{\pi r_2^2}{2}\right) = \frac{\pi}{2}(R^2 - r_1^2 - r_2^2)$.

However, $R = r_1 + r_2$; therefore, $\frac{\pi}{2}\left(R^2 - r_1^2 - r_2^2\right) = \frac{\pi}{2}\left((r_1 + r_2)^2 - r_1^2 - r_2^2\right)$ $= \pi r_1 r_2$.

The diameter, $FG$, of the circle with center, $S$, is $2\sqrt{r_1 r_2}$; consequently, the radius is then $\sqrt{r_1 r_2}$, and the area is $\pi r_1 r_2$, which is the same as the area of the arbelos, which we have calculated earlier.

As we show in Figure 3.55, there is yet another unexpected geometric experience in the arbelos, that is, that line segments $AH$ and $BH$ amazingly contain points $F$ and $G$, respectively — a surprising collinearity appears which should be appreciated as it was unanticipated.

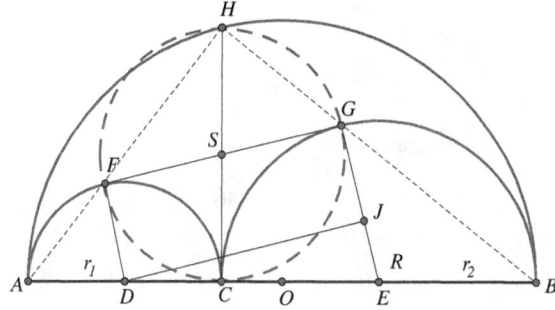

**Figure 3.55**

## Some More Entertainment

As a form of further entertainment, it might be nice to show an equality of areas. One such occurs when various circles intersect as shown in Figure 3.56. There we have two semicircles (with centers $D$ and $E$) along the diameter of the larger semicircle, one of which overlaps the larger semicircle, and one is below it. We then draw a tangent to the small semicircle at point $T$ from the external point $A$. With $AT$ as a diameter, we construct a circle with center $R$. As an entertainment, you may want to offer the following challenge, namely, to show that the area of the large circle with center $R$ is equal to the sum of the area of region resulting from the difference of the largest and smallest semicircles, plus the area of the bottom semicircle with center $D$.

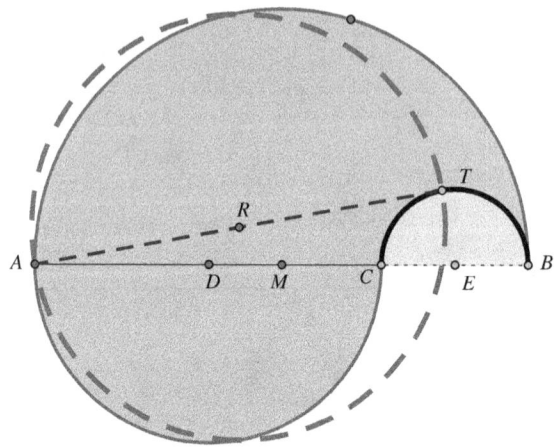

**Figure 3.56**

Although the last few entertainment options in geometry do require some recollection of high school mathematics, with a proper presentation they should be very accessible and be made interesting for the general audience.

We have now taken you through a wide variety of geometric entertainments, which typically engage the general audience because of its visual nature. However, as was said many times earlier, it is essential that the presenter offer the entertainments in a truly motivating fashion by showing genuine interest in what evolves.

## Endnotes

[1] Marcus Vitruvius Pollio was an ancient Roman writer, architect, and engineer.

[2] See Posamentier, Alfred. S. and Ingmar Lehmann. *The Glorious Golden Ratio*, Amherst, NY: Prometheus, Books, 2012.

[3] Since the perpendicular bisector of a chord, in this case *AB*, also bisects the arc *AB* at point *T*, it is rather simple to show that the four triangles in the quadrilateral *AOBT* are congruent. Therefore, the diagonals bisect each other, establishing *M* as the midpoint of *OT*.

[4] Honsberger, Ross, *Mathematical Delights*, Washington, DC: Mathematical Association of America, 2004. pp. 27–31.

# Chapter 4

# A Potpourri of Mathematical Entertainments

Mathematics can be made entertaining in many ways, unfortunately, this is often neglected in the school setting, where meeting curriculum standards overarches everything else. Teaching to the test is the worst anti-entertainment aspect of the instructional program. As we have seen so far, entertainment in mathematics can take on many variations. Yet, the delivery of these entertainment ideas — the enthusiasm of the presenter — is most essential. Curious mathematical challenges can frequently inspire an audience, especially if the solution is awe inspiring! Here we now provide some curiosities that should serve well to inspire your audience.

## Be Aware!

We are not trying to be tricky, but when playing with mathematics, one must keep all options open. For example, suppose you are given the number 316 and asked to rearrange the digits so that you create a number which is divisible by 7. Most people start to rearrange the number by reversing the digits and interchanging the digits and become a bit frustrated, since none of those result in a number which is divisible by 7. However, by being flexible and flipping one of the numbers, we can get the number 931, which is divisible by 7 (133 × 7 = 931). Remember, the original challenge was to rearrange the digits, which was done, albeit in a non-traditional fashion. This is the kind of thinking that sometimes needs to be done when presented with a mathematical challenge.

While we are on this way of thinking, consider correcting this equation without touching the numerals XI + I = X? The challenge presented here sounds absurd. How can you possibly change this equation to make it correct? The answer is simple. Look at it from the upside down and what you get is X = I + IX. This will once again open the minds of the audience to "think out of the box."

## A Cute Conundrum

Here is a nice little trick you can use to entertain an audience. Begin by selecting a person to work with you on this and have the person take a batch of coins in his right hand and in his left hand, with the proviso that in one hand there will be an even number of coins, while in the other hand there will be an odd number of coins. You are not supposed to know in which hand are the odd or the even number of coins. Ask your friend to multiply the number of coins in his right hand by 2, and the number of coins in his left hand by 3. Have your friend add the two products and tell you what the sum is. If the sum is even, then in the left hand had an even number of coins. If the sum is odd, then the left hand had an odd number of coins. Simple arithmetic analysis can be used to justify this trick.

## A Challenging Situation

Suppose you need to get exactly 4 pints of water from a stream and you only have two bottles at your disposal. One bottle has a capacity of 3 pints and the other bottle has a capacity of 5 pints. How can you measure out 4 pints of water to bring back from the stream? You should probably give your audience a few minutes time to try to figure out a proper solution.

One way of doing this is to fill the 5 pints bottle and pour off 3 pints by filling the second bottle, leaving 2 pints in the 5-pint bottle. Then empty the 3-pint bottle and pour the remaining 2 pints from the 5-pint bottle into the empty 3-pint bottle. Once again, fill the 5-pint bottle and then pour off enough water into the 3-pint bottle — which already contains 2 pints from the previous exchange. The amount of water poured off from the 5-pint bottle is 1 pint, thereby leaving 4 pints of water in the 5-pint bottle, which was your desired goal.

## Confusing or Simple?

If, on the average, a hen and a half can lay an egg and a half in a day and a half, how many eggs can six hens lay in eight days? This sounds somewhat confusing yet should prove to be entertaining and enlightening once the solution is produced.

There are various ways to solve this problem. Some might consider the following solution to be more sophisticated. Since $\frac{3}{2}$ hens work for $\frac{3}{2}$ days, we might consider this job to be $(\frac{3}{2})(\frac{3}{2}) = \frac{9}{4}$ "hen–days." In a similar fashion, the second job would be $6 \times 8 = 48$ "hen–days." By letting $x =$ the number of eggs laid by six hens and eight days, we can set up the following proportion: $\frac{\frac{9}{4}\,\text{hen–days}}{48\,\text{hen–days}} = \frac{\frac{3}{2}\,\text{eggs}}{x\,\text{eggs}}$, which then can be written as $\frac{9}{4}x = 48(\frac{3}{2}) = 72$, and then $x = 32$.

Another way of solving this problem would be to set up the following table:

| Number of hens laying eggs | Number of eggs produced | Number of days |
|---|---|---|
| $\frac{3}{2}$ | $\frac{3}{2}$ | $\frac{3}{2}$ |
| 3 | 3 | $\frac{3}{2}$ |
| 3 | 6 | 3 |
| 3 | 2 | 1 |
| 6 | 4 | 1 |
| 6 | **32** | 8 |

This problem will show how logic and algebra in each of the respective methods can solve a simple problem that looks complicated at the outset, but is not as challenging as it initially appeared.

## A Difficult-Appearing Problem Made Trivial

There are times when a problem appears rather difficult and yet with some experience or good insight it can be made rather trivial. Take, for example, the problem of finding the area of the shaded region in Figure 4.1, which is formed by two quarter circles inside square *ABCD*, whose side length is 1 unit.

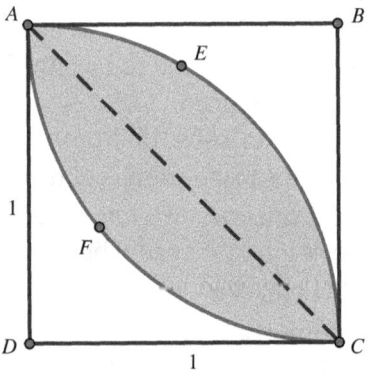

**Figure 4.1**

There are various ways to solve this problem. One common way would be to consider diagonal *AC* and find the area of the quarter circle *ADC*, which is $\frac{\pi}{4}$ and subtract the area of triangle *ADC*, which is $\frac{1}{2}$, to get $\frac{\pi}{4} - \frac{1}{2}$ and then doubling that segment to get the required area, $2(\frac{\pi}{4} - \frac{1}{2}) = \frac{\pi}{2} - 1$.

Perhaps a more elegant method would be to simply take the area of the quarter circle *ADCE*, which is $\frac{\pi}{4}$ and add it to the area of quarter circle *ABCF*, which is $\frac{\pi}{4}$ to get a total area of $\frac{\pi}{2}$ and then subtract the area of square *ABCD*, to get $\frac{\pi}{2} - 1$, which leaves the shaded region, as it was used twice as an overlap. The unusual strategy provides a rather elegant solution which could be enlightening.

## Fermat Extended

It is not too often that mathematics appears as a news item. However, on January 31, 1995,[1] the New York Times reported that "Fermat's Last Theorem" was finally proved by Andrew Wiles[2] after 358 years of lying dormant and unsolved. In the margin of one of his algebra books, Pierre de Fermat, a famous French mathematician (1607–1665), wrote that the equation $a^n + b^n = c^n$ has no integral solutions for $n > 2$. He mentions that there was not enough space in the margin of the book where he made the statement to prove this conjecture. To the present day we do not know whether he had a proof or not. However, it is entertaining to notice that if we extend this to $a^n + b^n + c^n = d^n$, a value for $n > 2$ can be found, namely, $n = 3$. In this case, $3^3 + 4^3 + 5^3 = 6^3$. This example merely shows how experiments on Fermat's conjecture over the years have exposed other analogous relationships, yet still leaving Fermat's conjecture as an open challenge. It is often nice to show how an event from the history of mathematics can provide interesting relationships and perhaps allow us to learn from them.

## Polygonal Sums

It can be entertaining to provide an audience with what seems to be a simple task but sometimes becomes a bit more challenging. Let's begin with a simpler version by using a triangle and placing the numbers in such a way that the sum of the four numbers along each side is the same. We offer two possibilities in Figure 4.2, but clearly, your audience may seek others.

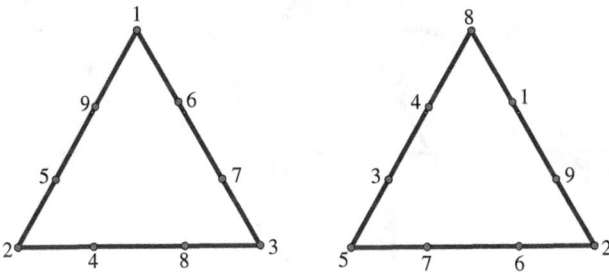

**Figure 4.2**

Another such puzzle can be shown on the pentagram, where the audience is asked to place the numbers from 1 to 12 at the intersections of a pentagram's sides so that the sum of all lines of 4 numbers is 24. Although there are 10 numbers to be used, we omitted the numbers 7 and 11, since it is impossible to perform this task with the numbers merely from 1 to 10. Naturally, the pentagram should be provided without the numbers that we have indicated in Figure 4.3.

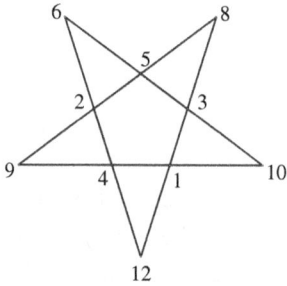

**Figure 4.3**

You might want to extend this challenge to another one similar to the previous one. Here you are to ask the audience to place the numbers from 1 to 12 along the intersections of the hexagram, as we did in Figure 4.4, in such a way that the sum along each line is 26 and the sum of the numbers at the 6 vertices of the hexagram is also 26. Furthermore, the sum of the numbers of the 6 vertices of the central hexagon is $2 \times 26 = 52$. In Figure 4.4, we offer three such possibilities.

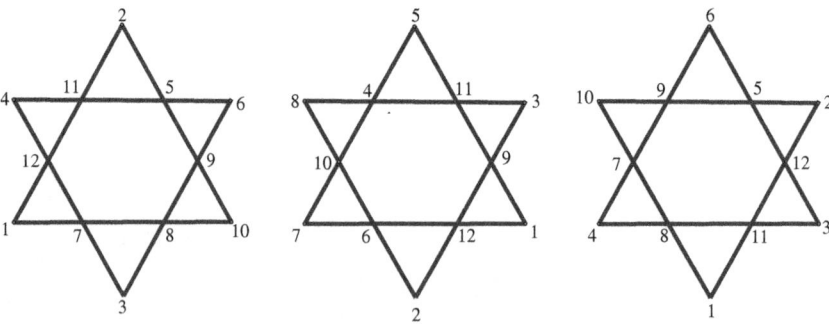

**Figure 4.4**

Now, if you really want to entertain your audience, have them find a way to have the sum of the numbers 1–6 placed at the circle intersections so that the sum along each circle is the same. We offer one solution in Figure 4.5.

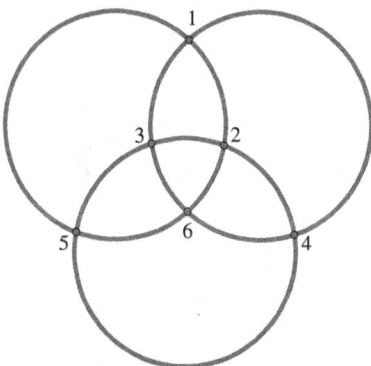

**Figure 4.5**

Once again we extend the challenge again. This time we have four mutually intersecting circles, and again the sum of the numbers at the circle intersections along each circle should be the same. In Figure 4.6, we provide a solution.

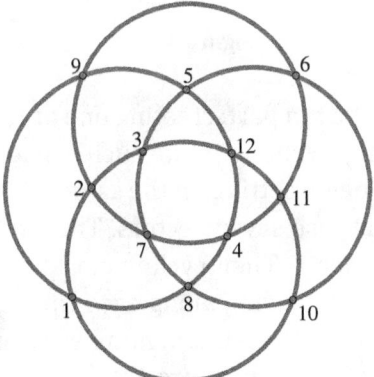

**Figure 4.6**

We conclude this delightful geometric arrangement of numbers with the following amazing circular offering which we show in Figure 4.7.

Here you will see that the sum of the numbers in each ring is 65, the sum of the numbers in each sector is 65 and if you create a counterclockwise spiral starting at the middle, such as the spiral 1, 18, 10, 22, 14, the total will be once again, amazingly, 65. This should surely amaze and ultimately entertain your audience!

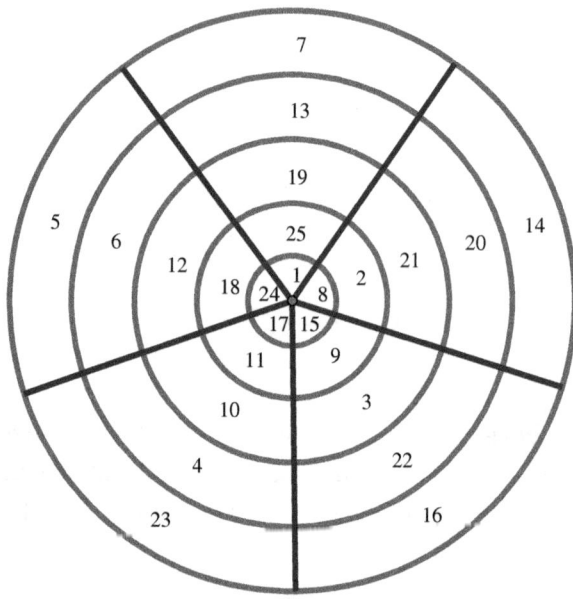

**Figure 4.7**

Now here comes the fun part: creating one of these "magic circles" from scratch. Suppose we choose to do such a magic circle with more than five sectors. We begin setting up the same odd number of sectors as of circles — in this case seven sectors. Then select any point and label it with the number 1. Then cycling clockwise and outward we place the number 2, and moving along each time clockwise and out-ward, you will notice the placement of numbers 3 and 4. Follow along as we create the magic circle in Figure 4.8.

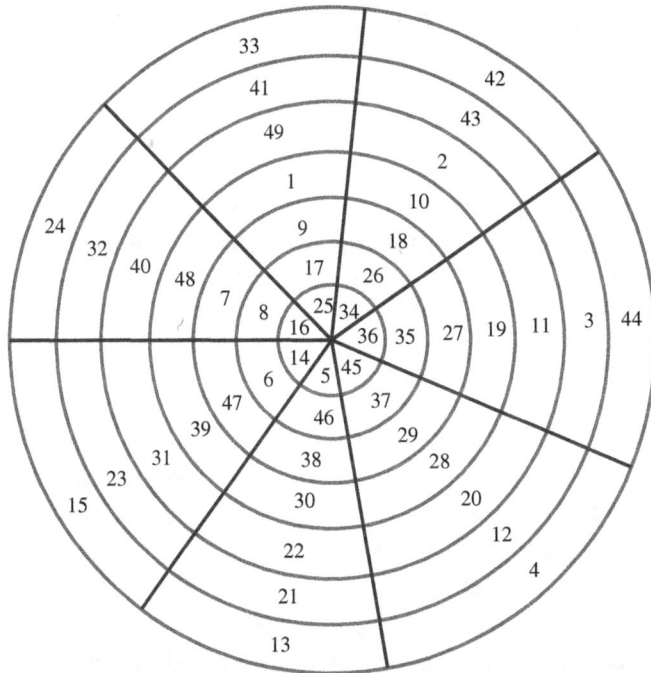

**Figure 4.8**

When you reach the outer ring, the next move should be to place next number (5) at the center spot of the next sector. Then continue on, again spiraling clockwise and outward. When the space to which you might be guided is occupied, make your next move in the same sector you were in, but one space closer to the center, which we have done where the number 8 was to go in the space occupied by the number 1, and we can place the 8 in the same sector as the 7, but closer to the center by one space. As we continue on, you will notice the spiral goes unimpeded till we get to the number 13, and we place the number 14 in the next sector at the center position. Now comes an unusual modification: since the space with a number 15 should occupy is already taken by the number 8, we place the number 15 in the same sector as the number 14, but at the outermost ring. The number 16 then goes to the center position in the next sector and the spiraling continues through the number 21. Since the space where the 22 should go is occupied, it simply goes to one space closer to the center in the same sector as the number 21. This process then continues along until you get to the last number, which in this case is the number 49. When you test the sums of the numbers in each of rings, you should find the same total of 175. Remember you can do this with other circles as long as the odd number of sectors is equal to the odd number of rings. Your audience should get a sense of accomplishment out of successfully completing a magic circle.

## The Difference Triangle

Challenges are often a source of entertainment — especially if the challenge is easy to understand. When simply asked to find the difference of numbers that should cause no difficulty of understanding. Here we introduce a game that will provide this sort of entertainment. This little game features 15 boxes as you can see in Figure 4.9. The challenge here is to be able to place the numbers from 1 to 15 in the boxes so that each number represents the difference of the two numbers below it. To be helpful, we will be inserting the first three numbers, so you can see that the two numbers below the number 5 have a difference equal to 5. Now try to do the rest of the numbers without looking further at the solution which we show in Figure 4.10.

**Figure 4.9**

Perhaps it is not easy not to look ahead, but if you present this to your audience make sure they don't see the solution which we provided in Figure 4.10.

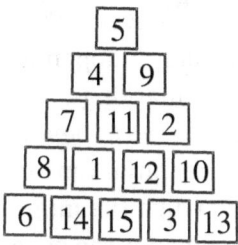

**Figure 4.10**

What we have done here with a five-row difference triangle, can also be done with a two-row, three-row or four-row triangles of this nature. However, it has been proved that it cannot be any larger than the triangle we show above with five rows. Your audience may wish to try to create other difference triangles.

## Having Fun with a Clock

The clock can be an interesting source of mathematical entertainment. Let's begin our journey through the face of a clock. We can begin by determining the *exact time* that the hands of a clock will overlap after 4:00. Your first reaction to the solution to this problem is likely to be that the answer is simply 4:20. But that does not take into account the hour hand's uniform movement while the minute hand also uniformly moves faster. With this in mind, the astute reader will begin to estimate the answer to be between 4:21 and 4:22. You should realize that the hour hand moves through any one of the intervals between minute markers every 12 minutes. Therefore, it will leave the interval 4:21–4:22 at 4:24. This, however, does not answer the original question about the *exact time* that the overlap of the two hands takes place.

We will provide a "trick" to better deal with this situation. Realizing that this first guess of 4:20 is not the correct answer, since the hour hand does not remain stationary and moves when the minute hand moves, the trick is to simply multiply the 20 (the wrong answer) by $\frac{12}{11}$ to get $21\frac{9}{11}$, which yields the correct answer: $4:21\frac{9}{11}$.

One way to understand the movement of the hands of a clock is by considering the hands traveling independently around the clock at uniform speeds. The minute markings on the clock (from now on referred to as "markers") will serve to denote distance as well as time. Now consider 4 o'clock as the initial time on the clock. Our problem will be to determine exactly when the minute hand will overtake the hour hand after 4 o'clock. Consider the speed of the hour hand to be $r$, then the speed of the minute hand must be $12r$. We seek the distance, measured by the number of markers traveled that the minute hand must travel to overtake the hour hand. Let us refer to this distance as $d$ markers. Hence the distance that the hour hand travels is $d - 20$ markers, since it has a 20-marker head start over the minute hand (see Figure 4.11).

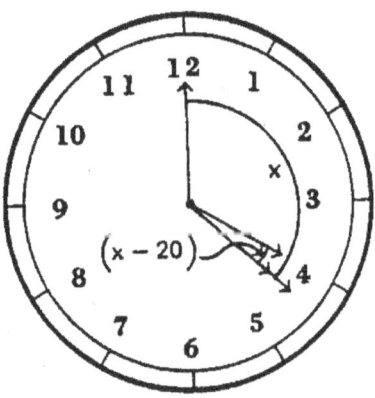

**Figure 4.11**

For this to take place, the times required for the minute hand, $\frac{d}{12r}$, and for the hour hand, $\frac{d-20}{r}$, are the same. Therefore, $\frac{d}{12r} = \frac{d-20}{r}$, and $d = \frac{12}{11} \cdot 20 = 21\frac{9}{11}$. Thus, the minute hand will overtake the hour hand at exactly $4:21\frac{9}{11}$.

Consider the expression $d = \frac{12}{11} 20$. The quantity 20 is the number of markers that the minute hand had to travel to get to the desired position, if we assume the hour hand remained stationary. However, quite obviously, the hour hand does not remain stationary. Hence, we must multiply this quantity by $\frac{12}{11}$, since the minute hand must travel $\frac{12}{11}$ as far. We can, therefore, refer to this fraction, $\frac{12}{11}$, as the "correction factor."

To become more familiar with the use of the correction factor, we shall choose some short and simple problems. For example, you may seek to find the exact time when the hands of a clock overlap between 7 and 8 o'clock. Here, you would first determine how far the minute hand would have to travel from the "12" position to the position of the hour hand, assuming again that the hour hand remains stationary. Then by multiplying this number of markers, 35, by the correction factor, $\frac{12}{11}$, you will obtain the exact time, $7:38\frac{2}{11}$, that the hands will overlap.

To enhance your understanding of this new procedure, consider a person checking a wristwatch against an electric clock and noticing that the hands on the wristwatch overlap every 65 minutes (as measured by the electric clock). Is the wristwatch fast, slow, or accurate?

You may wish to consider the problem in the following way. At 12 o'clock the hands of a clock overlap exactly. Using the previously described method, we find that the hands will again overlap at exactly $1:05\frac{5}{11}$, and then again at exactly $2:10\frac{10}{11}$, and again at exactly $3:16\frac{4}{11}$, and so on. Each time there is an interval of $65\frac{5}{11}$ minutes between overlapping positions. Hence, the person's watch is inaccurate by $\frac{5}{11}$ of a minute. Can you now determine if the wristwatch is fast or slow?

There are many other interesting, and sometimes rather difficult, problems made simple by this "correction factor." You may very easily pose your own problems. For example, you may wish to find the exact times when the hands of a clock will be perpendicular (or form a straight angle) between, say, 8 and 9 o'clock.

Again, you would try to determine the number of markers that the minute hand would have to travel from the "12" position until it forms the desired angle with the stationary hour hand. Then multiply this number by the correction factor, $\frac{12}{11}$, to obtain the exact actual time. That is, to find the exact time when the hands of a clock are *first* perpendicular to each other between 8 and 9 o'clock, determine the desired position of the minute hand when the hour hand remains stationary (that would be on the 25-minute marker). Then, multiply

25 by $\frac{12}{11}$ to get $8:27\frac{3}{11}$, the exact time when the hands are *first* perpendicular after 8 o'clock.

For those who want to look at this issue from a non-algebraic viewpoint, you could justify the $\frac{12}{11}$ correction factor for the interval between overlaps in the following way.

Think of the hands of a clock at noon. During the next 12 hours (i.e., until the hands reach the same position at midnight), the hour hand makes one revolution, the minute hand makes 12 revolutions, and the minute hand coincides with the hour hand 11 times (including midnight, but not noon, starting just after the hands separate at noon). Since each hand rotates at a uniform rate, the hands overlap each $\frac{12}{11}$ of an hour, or $65\frac{5}{11}$ minutes. This can be extended to other situations. You should derive a great sense of achievement and enjoyment as a result of employing this simple procedure — with our correction factor — to solve what usually appears to be a very difficult clock problem.

## What is Relativity?

The concept of relativity is generally not well understood by most non-scientists. Although it is often associated with Albert Einstein, it has many applications. It may be a difficult concept to grasp for some, so we shall exercise patience and support as we gently navigate further, and present the way to make this concept intelligible. Consider the following problem:

> *While rowing his boat upstream, David drops a cork overboard and continues rowing for 10 more minutes. He then turns around, chasing the cork, and retrieves it when the cork has traveled one mile downstream. What is the rate of the stream?*

Rather than approach this problem through the traditional methods, common in an algebra course, let us consider the following. The problem can be made significantly easier by considering the notion of relativity. It does not matter if the stream is moving and carrying David downstream, or is still. We are concerned only with the separation and coming together of David and the cork. If the stream were stationary, David would require as much time rowing to the cork as he did rowing away from the cork. That is, he would require 10 + 10 = 20 minutes. Since the cork travels one mile during these 20 minutes, the stream's rate of speed is 3 miles per hour. Again, this may not be an easy concept to grasp for some and is best left to ponder in quiet. It is a concept worth understanding, for it has many useful applications in the everyday life thinking processes. This is, after all, one of the purposes for understanding mathematics, beyond providing us an opportunity for some entertainment.

## Successive Percentages

Most folks find percentage problems to have long been an unpleasant nemesis. Problems get particularly confusing when multiple percentages need to be processed in the same situation. This presentation — as entertaining it might be — can turn this one-time nemesis into a delightfully simple arithmetic algorithm that affords lots of useful applications and provides new insight into successive-percentage problems. This not-very-well-known procedure should enchant your audience. Let's begin by considering the following problem:

> *Wanting to buy a coat, Barbara is faced with a dilemma. Two competing stores next to each other carry the same brand coat with the same list price, but with two different discount offers. Store A offers a 10% discount year-round on all its goods, but on this particular day, Store A offers an additional 20% on top of their already discounted price. Store B simply offers a discount of 30% on that day in order to stay competitive. Which store offers the better deal? How many percentage points difference, if any, is there between the two options open to Barbara?*

At first glance, you may assume there is no difference in price, since you may feel that 10% + 20% = 30%, yielding the same discount in both cases. Yet, with a little more thought you may realize that this is *not* correct, since in store A only 10% is calculated on the original list price, with the 20% calculated on the lower price, while at store B, the entire 30% is calculated on the original price. Now, the question to be answered is, what percentage difference is there between the discounts offered in store A and store B?

One common procedure might be to assume the cost of the coat to be $100, calculate the 10% discount yielding a $90 price, and an additional 20% of the $90 price (or $18) will bring the price down to $72. In store B, the 30% discount on $100 would bring the price down to $70, giving a discount difference of $2, which in this case is 2%. This procedure, although correct and not too difficult, is a bit cumbersome and does not always allow a full insight into the situation.

An interesting and quite unusual procedure[3] is provided here for entertainment and fresh look into this problem situation. We offer a mechanical method for obtaining a single percentage discount (or increase) equivalent to two (or more) *successive discounts* (or increases).

(1) **Change each of the percents involved into decimal form:**

For the above problem we would get 0.10 and 0.20.

(2) **Subtract each of these decimals from 1.00:**

Therefore, our next step is to obtain 0.80 and 0.90 (for an increase, add to 1.00).

(3) **Multiply these differences:**

We then obtain $(0.80)(0.90) = 0.72$.

(4) **Subtract this number from 1.00:**

This gives us $1.00 - 0.72 = 0.28$, which represents the combined *discount.*

**(If the result of step 3 is greater than 1.00, subtract 1.00 from it to obtain the percent of *increase*.) When we convert 0.28 back to percent form, we obtain 28%, the equivalent of successive discounts of 20% and 10%.**

To answer the original question posed, the combined percentage of 28% differs from 30% by 2%.

Following the same procedure, you can also combine more than 2 successive discounts. Furthermore, successive increases, combined or not combined with a discount, can also be accommodated in this procedure by adding the decimal equivalent of the increase to 1.00, where the discount was subtracted from 1.00 and then continue the procedure in the same way. If the end result is greater than 1.00, then this end result reflects an overall increase rather than the discount as found in the above problem.

This procedure not only streamlines a typically cumbersome situation, but also provides some insight into the overall picture of percentages. For example, the question "Is it advantageous to the buyer in the above problem to receive a 20% discount and then a 10% discount, or the reverse, 10% discount and then a 20% discount?" The answer to this question is not immediately intuitively obvious. Yet, since the procedure just presented shows that the calculation is merely multiplication, a commutative operation, we find immediately that there is no difference between the two options.

So here you have a delightful algorithm for combining successive discounts or increases or combinations of these. Not only is it useful but also it gives your audience some new-found power in dealing with percentages, when using a calculator is not accessible.

## A Percentage Conundrum

It can be entertaining to provide your audience with an everyday problem, where the solution may be somewhat evasive. Consider the situation where Charlie bought 2 tablets and sold one to Max for $120 and made a profit of 25%. He sold the other tablet to Sam for $120 and had a loss of 25%. Usually people will think that he broke even. What might your audience think at this point?

Let's consider the first tablet, which he sold for $120 and made a profit of 25%, which would indicate that he paid $96, since he made a $24 profit. Selling the other tablet to Sam, where he had a loss of

25%, he paid $160, which is a loss of $40. Therefore, he lost 24 − 40 = 16, which is the of loss − $16 in the entire transaction. Initially, this is an unexpected or perhaps even counterintuitive solution, however, it will give the audience useful insight and perhaps some entertainment as well.

## Let Algebra Help Reasoning

There are times when the solution to a problem is rather counterintuitive. A simple problem such as the following leaves one wondering "why couldn't I have logically come to that conclusion?"
Here's the problem:

> *The cost of a can of tuna fish is 50¢, and the cost of the tuna fish is 30¢ more than the can. What is the cost of the can?*

Typically, the first reaction would be the cost of the can should be 20¢. However, with the help of algebra, we can solve the problem rather simply. If we let $C$ equal the cost of the can and let $T$ equal the cost of the tuna fish. Therefore, the cost of the can of tuna fish is $C + T = 50$, and the cost of the tuna fish is $T = C + 30$. When we substitute for $T$ in the first equation, we get $C + (C + 30) = 50$. Therefore, $C = 10$, which is not the anticipated result as was indicated earlier.

## Another Counterintuitive Situation to Ponder

We have available a series of three books each of which is 2 inches wide and they are placed side-by-side in the order of Volumes I–III. A bookworm bores a hole from the front cover of Volume I and reaches the back cover of Volume III. The question is how far has the bookworm traveled?

Naturally, the expected answer is that the bookworm will have traveled 6 inches, that is, through each of the books. However, if you look at Figure 4.12, you will see that the bookworm will only have traveled 2 inches, since he will have merely traveled through Volume II.

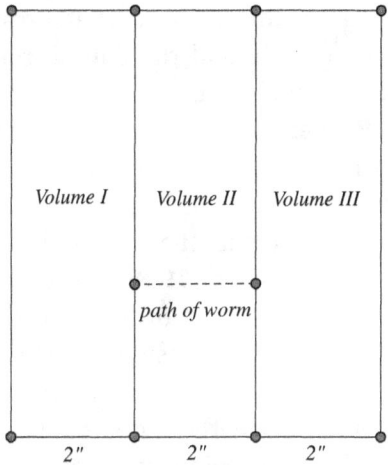

**Figure 4.12**

# Coloring a Map

Begin by asking your audience: Have you ever wondered how a map is colored? Aside from deciding which colors should be used, the question might come up regarding the number of colors that are required for a specific map so that there will never be the same color on both sides of a boarder. Well, mathematicians have determined the answer to this question, namely, that one will never need more than four colors to color any map, regardless of how many borders or contorted arrangements the map presents. For many years, the question as to how many colors are required was a constant challenge to mathematicians, especially those doing research in topology, which is a branch of mathematics related to geometry, where figures discussed may appear on plane surfaces or on three-dimensional surfaces. The topologist studies the properties of a figure that remain the same *after* the figure has been distorted or stretched according to a set of rules. A piece of string with its ends connected may take on the shape of a circle, or a square, which is all the same for the topologist. In going through this transformation, the order of the "points" along the string does not change. This retention of ordering has survived the distortion of shape, and it is this property that attracts the interest of topologists. Therefore, a circle and the square represent the same geometric concept to the topologist.

Throughout the 19th century, it was believed that five colors were required to color even the most complicated looking map. However, there was always strong speculation that four colors would suffice. It was not until 1976 that the mathematicians Kenneth Appel (1932–2013) and Wolfgang Haken (1928–) "proved" that four colors were sufficient to color any map. However, unconventionally they used a high-powered computer to consider all possible map arrangements. It must be said that there are still mathematicians who are dissatisfied with the proof, since it was done by computer and not in the traditional way "by hand." Previously, it was considered one of the famous unsolved problems of mathematics. Let us now delve into the consideration of various maps and the number of colors required to color them in such a way that no common boundary of two regions shares the same color on both of the sides. Your audience may also want to conjure up a difficult map as an example to consider. This is clearly a requirement for coloring any map.

Suppose we consider a geographic map that has a configuration analogous to that shown in Figure 4.13.

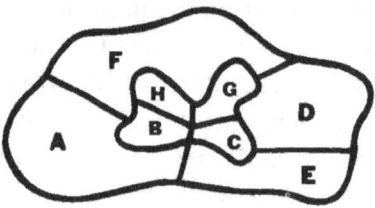

**Figure 4.13**

Here we notice that there are eight different regions indicated by the letters shown. Suppose we list all regions that have a common boundary with region $H$, and regions that share a common vertex with region $H$. The regions designated by the letters $B$, $G$, and $F$ share a border with region $H$. The region designated by the letter $C$ shares a vertex with the region designated by the letter $H$.

Remember, a map will be considered correctly colored when all the regions are completely colored and any two regions that share a common boundary have different colors. Two regions sharing a common vertex may also share the same color. Let's consider coloring a few maps (Figure 4.14) to see various configurations a map can have that requires not more than three colors. (***b***/blue; ***r***/red; ***y***/yellow; ***g***/green).

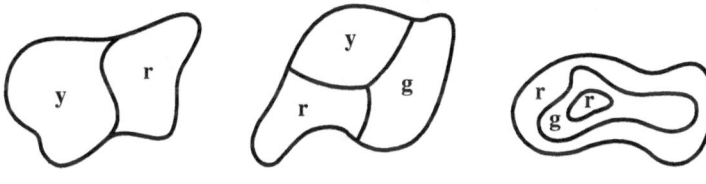

**Figure 4.14**

The first map is able to be colored in two colors: yellow and red. The second map required three colors: yellow, red, and green. The third map has three separate regions, but only requires two colors, red and green, since the innermost territory does not share a common border with the outermost territory.

It would seem reasonable to conclude that if a three-region map can be colored with less than three colors, a four-region map can be colored with less than four colors. Let's consider such a map.

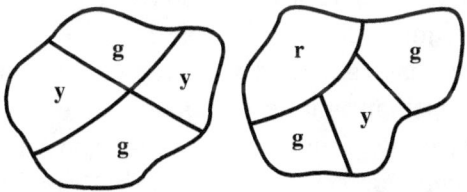

**Figure 4.15**

The left-side map shown in Figure 4.15 has four regions and requires only two colors for correct coloring. Whereas, the right-side map in Figure 4.15 also consists of four regions, but requires three colors for correct coloring.

We should now consider a map that requires four colors for proper coloring of the regions. Essentially, this will be a map, where each of the four regions shares a common border with the other three regions. One possible such mapping is shown in Figure 4.16.

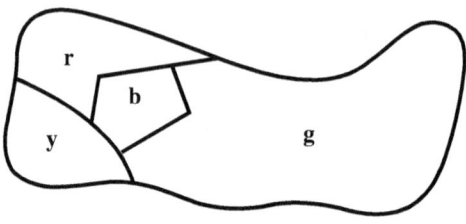

**Figure 4.16**

If we now take the next logical step in this series of map-coloring challenges, we should come up with the idea of coloring maps involving five distinct regions. It will be possible to draw maps that have five regions, which require two, three, or four colors to be colored correctly. The task of drawing a five-region map that *requires* five colors for correct coloring will be impossible. This curiosity can be generalized through further investigation and should convince you that the idea that any map, on a plane surface, with any number of regions, can be successfully colored with four or fewer colors. You might want to challenge friends to conceive of a map of any number of regions that requires more than four colors so that no two regions with a common border share the same color.

This challenge has remained alive for many years, challenging some of the most brilliant minds, but as we said earlier, the issue has been closed by the work of the two mathematicians, Appel and Haken. There are still many conjectures in mathematics that have escaped proof, but have never been disproven. Here, at least, we have a conjecture that is closed. Such a story is also a source of entertainment for the general audience.

# Crossing the Bridges

New York City, which is comprised mostly of islands (as a matter of fact the only part of the city that is part of the mainland of the United States is the Bronx — of course, excluding City Island), is blessed with many bridges. We have bike and track races that traverse several bridges throughout their tour. One takes bridge crossing for granted these days. They essentially become part of the path traveled and don't become noticed unless there is a toll to be paid. Then one takes particular note of the bridge crossing. In the 18th century and earlier, when walking was the dominant form of local transportation, people would often count particular kinds of objects they passed. One such was bridges. Through the 18th century, the small Prussian city of Königsberg (today called Kaliningrad, Russia), located where the Pregel River forms two branches, was faced with a recreational dilemma: Could a person walk over each of the seven bridges *exactly once* in a continuous walk through the city? The residents of the city had this as a recreational challenge, particularly on Sunday afternoons. Since there were no successful attempts, the challenge continued for many years.

This problem, although entertaining by its very nature, provides a wonderful window into networks, which is referred to as graph theory, an extended field of geometry, which gives us a renewed view of the subject. To begin we should present the problem. In Figure 4.17, we can see the map of the city with the seven bridges highlighted.

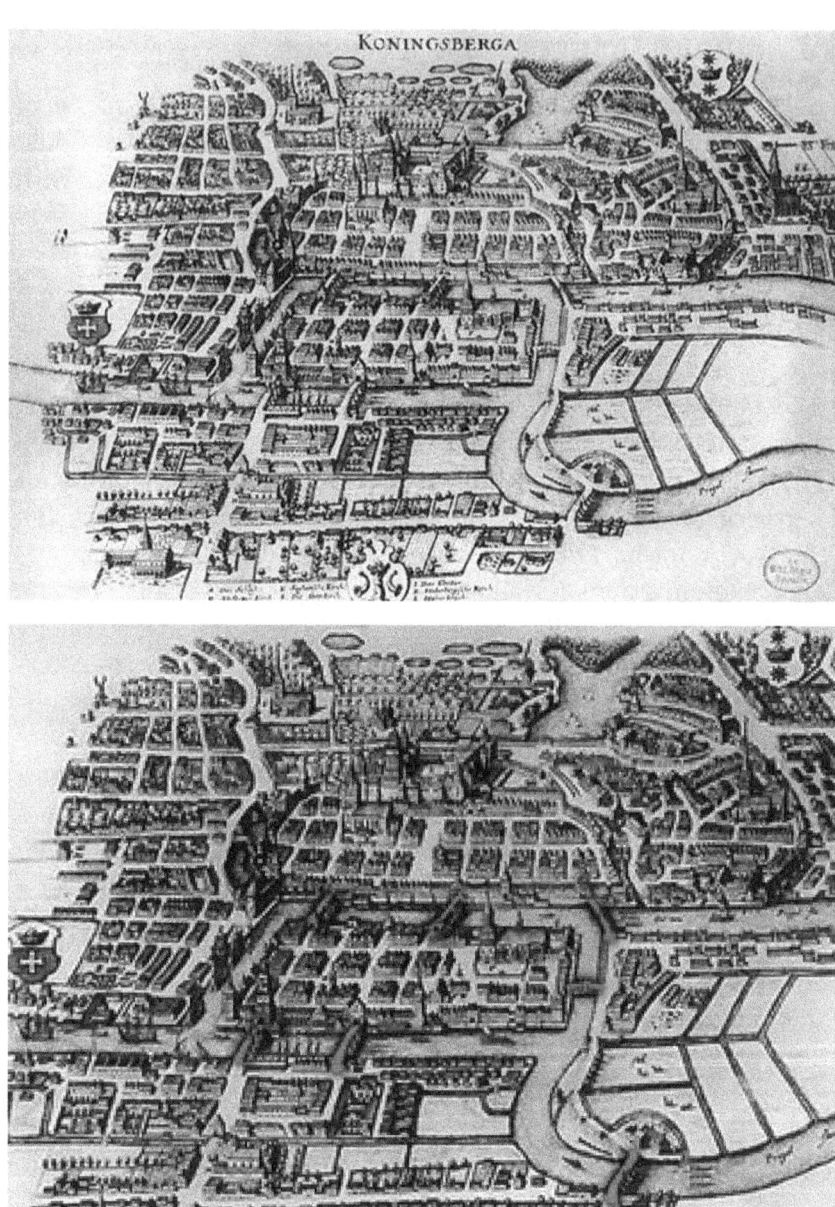

**Figure 4.17**

In Figure 4.18, we will indicate the island by A, the left bank of the river by B, the right bank by C, and the area between the two arms of the upper course by D. If we start at Holz and walk to Schmiede and then through Honig, through Hohe, through Köttel, through Grüne, we will never cross Krämer. On the other hand, if we start at Krämer and walk to Honig, through Hohe, through Köttel, through Schmiede, through Holz, we will never travel through Grüne.

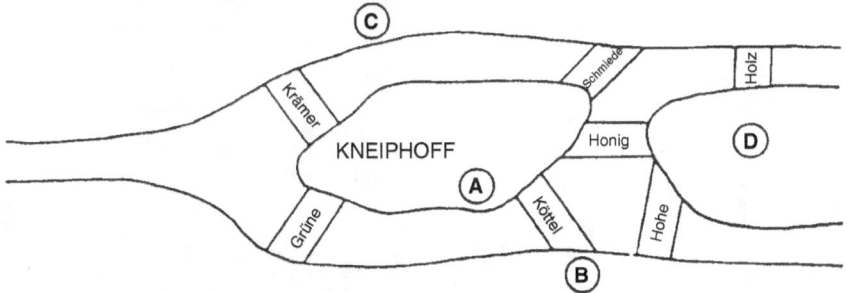

**Figure 4.18**

In 1735, the famous Swiss mathematician Leonhard Euler (1707–1783) proved mathematically that this walk could not be performed. The famous Königsberg Bridges Problem, as it became known, is a lovely application of a topological problem with networks. It is very nice to observe how mathematics used properly can put a practical problem to rest. Before we embark on the problem, we ought to become familiar with the basic concept involved. Toward that end, try to trace with a pencil each of the following configurations without missing any part and without going over any part twice. Keep count of the number of arcs or line segments, which have an endpoint at each of the points: *A, B, C, D, E*.

Configurations, called networks, such as the five figures shown in Figure 4.19 are made up of line segments and/or continuous arcs. The number of arcs or line segments that have an endpoint at a particular vertex is called the *degree* of the vertex.

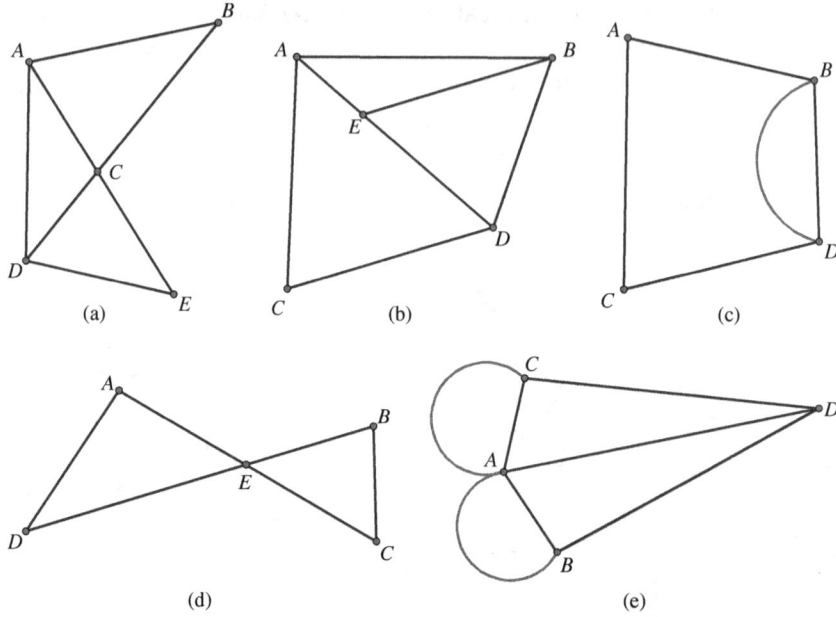

**Figure 4.19**

After trying to trace these networks without lifting the pencil off the paper and without going over any line more than once, you should notice two direct outcomes. The networks can be traced (or traversed) if they have (1) all even degree vertices or (2) exactly two odd degree vertices. The following two statements summarize this finding:

(1) There is an even number of odd degree vertices in a connected network.

(2) A connected network can be traversed, only if it has at most two odd degree vertices.

Network Figure 4.19(a) has five vertices. Vertices *B, C, E* are of even degree and vertices *A* and *D* are of odd degree. Since Figure 4.19(a) has exactly two odd degree vertices as well as three even degree vertices, it is traversable. If we start at A then go down to $D$, across to $E$, back up to $A$, across to $B$, and down to $D$, we have chosen a desired route.

Network Figure 4.19(b) has five vertices. Vertex $C$ is the only even degree vertex. Vertices $A$, $B$, $E$, and $D$ are all of odd degree. Consequently, since the network has more than two odd vertices, it is not traversable.

Network Figure 4.19(c) is traversable because it has two even vertices and exactly two odd degree vertices.

Network Figure 4.19(d) has five even degree vertices and, therefore, can be traversed.

Network Figure 4.19(e) has four odd degree vertices and *cannot* be traversed.

The Königsberg Bridge Problem is the same problem as the one posed in Figure 4.19(e). Let's take a look at Figures 4.19(e) and 4.18, and note the similarity. There are seven bridges in Figure 4.18 and there are seven lines in Figure 4.19(e). In Figure 4.19(e), each vertex is of odd degree. In Figure 4.18, if we start at $D$ we have three choices we could go to Hohe, Honig, or Holz. If in Figure 4.19(e) we start at $D$, we have three line paths to choose from. In both figures, if we are at $C$ we have either three bridges we could go on (or three lines). A similar situation exists for locations $A$ and $B$ in Figure 4.18 and vertices $A$ and $B$ in Figure 4.19(e). We can see that this network cannot be traversed.

By reducing the bridges and islands to a network problem we can easily solve it. This is a clever tactic to solve problems in mathematics. You might want to have members of the audience try to find a group of local bridges to create a similar challenge and see if the walk is traversable. This problem and its network application is an excellent introduction into the field of topology.

We can also apply this technique of the traversability of a network to the famous *five-bedroom-house* problem. Let's consider the floor plan of a house with five rooms as shown in Figure 4.20.

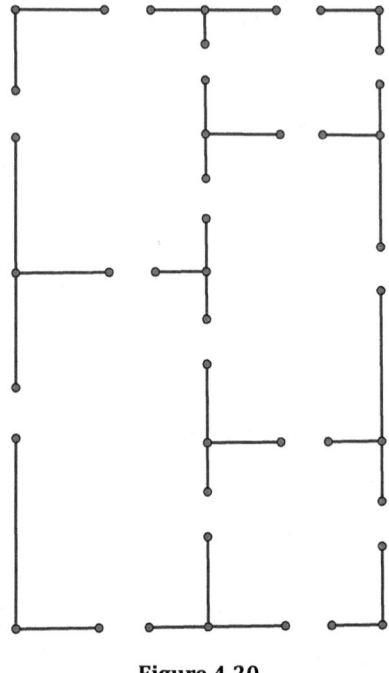

**Figure 4.20**

Each room has a doorway to each adjacent room and a doorway leading outside the house. The problem is to have a person start either inside or outside the house and walk through each doorway exactly once. You will realize that, although the number of attempts is finite, there are far too many ways to make a trial-and-error solution practical. Figure 4.21 shows various possible paths joining the five rooms A, B, C, D, and E and the outside area F.

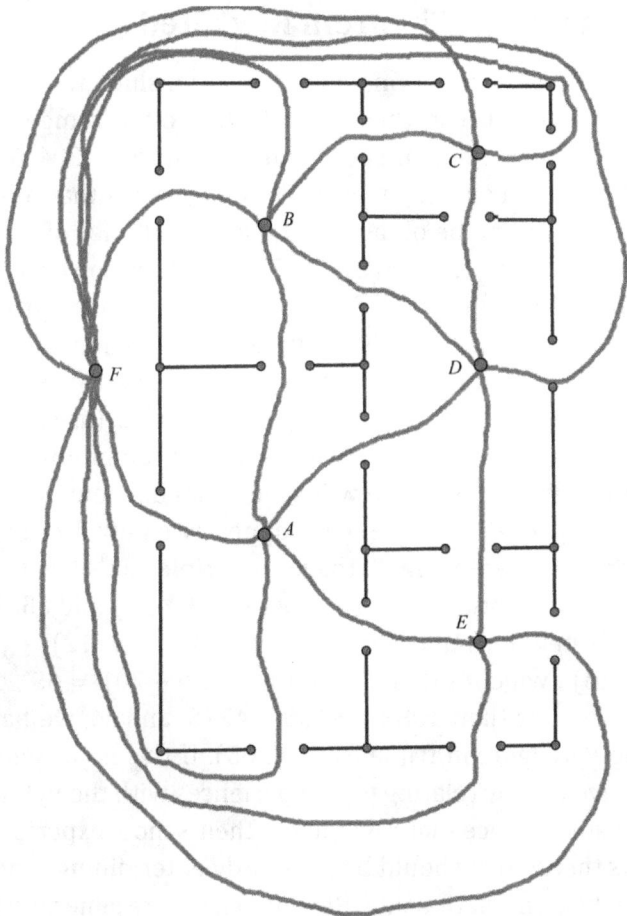

**Figure 4.21**

As before, our question can be answered by merely determining if this network is traversable. In Figure 4.21, we notice that we have marked the vertices with the letters *A, B, C, D, E,* and *F*. We notice that 4 vertices are of odd degree, and 2 vertices are of even degree. Since there are not exactly 2 or 0 vertices of odd order, this network cannot be traversed. Therefore, the five-room-house problem does not have a solution path, which would allow walking through each doorway exactly once. As your audience can see, even in choosing paths for travel, mathematics seems to provide a solution to our question.

## The Pythagorean Theorem Revisited

Perhaps, one of the most remembered relationships taught in high school is the Pythagorean theorem, which is often remembered as $a^2 + b^2 = c^2$. The most popular such arrangements are: (3, 4, 5), (5, 12, 13), (7, 24, 25), and so on. Some readers may recall that other such triples of numbers can be obtained by taking multiples of these such as (6, 8, 10) or (10, 24, 26), or (21, 72, 75). There is, however, a rather little-known relationship that can also produce other Pythagorean triples which are not necessarily multiples of the original one and that can be done by combining two known Pythagorean triples in the following way. Let's consider to Pythagorean triples $(a, b, c)$, where $a^2 + b^2 = c^2$, and $(A, B, C)$, where $A^2 + B^2 = C^2$. By combining these two triples in the following way, we will have created a new Pythagorean triple: $(aA - bB)^2 + (aB + bA)^2 = (cC)^2$. Suppose we now use this procedure for trying to create new Pythagorean triple. Let's try it with the first two triples mentioned above, namely (3, 4, 5) and (5, 12, 13). Applying this new formula (above) we get $(3 \cdot 5 - 4 \cdot 12)^2 + (3 \cdot 12 + 4 \cdot 5)^2 = (5 \cdot 13)^2$, which is then $(15 - 48)^2 + (36 + 20)^2 = 65^2$, or $33^2 + 56^2 = 65^2$, which is then $1089 + 3136 = 4225$, and so, we have then created the Pythagorean triple (33, 56, 65). If this is presented in a very attractive way by relating past experiences with the Pythagorean theorem to an audience that remembers their school experience with this famous theorem, it should be pleasantly entertaining, since it is a rather little-known relationship. Remember, we are generating primitive Pythagorean triples, since they are not a multiple of a previously determined primitive triple. Your audience now has something very special for their knowledge chest.

## Endnotes

[1] Kolata, G. *How a Gap in the Fermat Proof Was Bridged.* New York Times, January 31, 1995.

[2] Posamentier, A. S. and C. Spreitzer. *Math Makers: The Lives and Works of 50 Famous Mathematicians.* Guilford, CT: Prometheus Books, 2020.

[3] It is provided without justification of its validity so as not to detract from the solution of the problem. However, for further discussion of this procedure, the reader is referred to A. S. Posamentier and B. S. Smith, *Teaching Secondary School Mathematics: Techniques and Enrichment Units* (World Scientific Publishing. 10th Edition, 2020).

# Final Thoughts

Throughout this book we have journeyed through a variety of topics that would demonstrate that mathematics, is not only one of the most important subjects in the school curriculum, but can also be quite entertaining. Naturally, the way one presents these topics to others is of key importance. If they are presented merely as a clever finding, lots could be lost from the enjoyment. Therefore, the well-prepared presentation can possibly be as important as the content. The old adage, "the tone makes the music," could be applied here as well. Furthermore, not only is the style of presentation important but also the audience appropriateness of the selected material must be carefully chosen. A fair number of the topics would be universally interesting, whereas others might be more for specific audiences. Your task as the reader is now to select which topics to bring "to the table" and for which audience they are best suited. Moreover, the order and timing of the presentation needs to be considered, and, by all means, ultimately presented with a great deal of enthusiasm. Typically, not only will the audience be entertained and impressed with the many curiosities you encountered throughout this book, but you as the presenter should also enjoy the presentation of these topics. Good luck, and enjoy mathematics for its power and beauty!

CPSIA information can be obtained
at www.ICGtesting.com
Printed in the USA
FSHW021826040820
72680FS

9 789811 219283